Springer Monographs in Mathematics

For further volumes:
www.springer.com/series/3733

Sourav Chatterjee

Superconcentration and Related Topics

 Springer

Sourav Chatterjee
Department of Statistics
Stanford University
Stanford, CA, USA

ISSN 1439-7382 ISSN 2196-9922 (electronic)
Springer Monographs in Mathematics
ISBN 978-3-319-03885-8 ISBN 978-3-319-03886-5 (eBook)
DOI 10.1007/978-3-319-03886-5
Springer Cham Heidelberg New York Dordrecht London

Library of Congress Control Number: 2014930163

Mathematics Subject Classification (2010): 60E15, 60K35, 60G15, 82B44, 60G60, 60G70

© Springer International Publishing Switzerland 2014
This work is subject to copyright. All rights are reserved by the Publisher, whether the whole or part of the material is concerned, specifically the rights of translation, reprinting, reuse of illustrations, recitation, broadcasting, reproduction on microfilms or in any other physical way, and transmission or information storage and retrieval, electronic adaptation, computer software, or by similar or dissimilar methodology now known or hereafter developed. Exempted from this legal reservation are brief excerpts in connection with reviews or scholarly analysis or material supplied specifically for the purpose of being entered and executed on a computer system, for exclusive use by the purchaser of the work. Duplication of this publication or parts thereof is permitted only under the provisions of the Copyright Law of the Publisher's location, in its current version, and permission for use must always be obtained from Springer. Permissions for use may be obtained through RightsLink at the Copyright Clearance Center. Violations are liable to prosecution under the respective Copyright Law.
The use of general descriptive names, registered names, trademarks, service marks, etc. in this publication does not imply, even in the absence of a specific statement, that such names are exempt from the relevant protective laws and regulations and therefore free for general use.
While the advice and information in this book are believed to be true and accurate at the date of publication, neither the authors nor the editors nor the publisher can accept any legal responsibility for any errors or omissions that may be made. The publisher makes no warranty, express or implied, with respect to the material contained herein.

Printed on acid-free paper

Springer is part of Springer Science+Business Media (www.springer.com)

Preface

Understanding the fluctuations of random objects is one of the major goals of probability theory. There is a whole subfield of probability and analysis, called concentration of measure, devoted to understanding fluctuations of random objects. Measure concentration has seen tremendous progress in the last forty years. And yet, there is a large class of problems in which classical concentration of measure gives suboptimal bounds on the order of fluctuations.

In 2008 and 2009, I posted two preprints on arXiv where it was shown that the suboptimality of classical concentration, when it occurs, is not simply a question of mathematical inadequacy. The suboptimality is in fact equivalent to a number of very interesting things going on in the structure of the random object under investigation. Indeed, the consequences are possibly interesting enough for the suboptimality of classical concentration to deserve a name of its own; I call it 'superconcentration'.

This monograph is a combination of these two preprints (which will not be published individually), together with some new material and new insights. The majority of the results are the same as in the preprints, but the presentation is radically different. In particular, I think I achieved a substantial degree of simplification and clarity through the use of the spectral approach. This is quite standard in the noise-sensitivity literature (which is intimately connected with the topic of this monograph), but it is not the way I derived the results in the preprints.

In addition to the theorems and proofs, I have interspersed the document with a sizable number of open problems for professional mathematicians and exercises for graduate students.

I spent many hours deliberating over whether to keep the book in its present form, or expand it to around 250 pages by including additional material from classical concentration of measure and other related topics. In the end I decided not to expand. The rationale behind this decision is two-fold: The first reason is that there are several comprehensive texts on concentration of measure already available in the market, and I did not wish to encroach on that territory. I had originally intended this book to be a short and succinct exposition of the superconcentration phenomenon, and in the end I decided to keep it that way. The second reason is that I am deeply familiar with my procrastinating tendencies, which made me confident that I would never have finished this project if I had planned a major overhaul. However I did expand a little bit; the original version that I submitted to Springer was even thinner. On the advice of one of the reviewers, I decided to include several additional examples that I had omitted in the first draft.

The main body of this monograph grew out of a set of six lectures I gave at the Cornell Probability Summer School in July 2012. The task of explaining the results and proofs to graduate students forced me to organize the material in a manner suitable for exposition in a monograph. I thank the organizers of CPSS 2012 for giving me this opportunity, and one of the students attending the summer school, Mihai Nica, for doing a terrific job in taking notes and typing them up.

I am grateful to Persi Diaconis for suggesting that I write this monograph, and for his constant encouragement and advice. I thank the reviewers for many useful comments, and Christophe Garban, Susan Holmes, Dmitry Panchenko and Michel Talagrand for looking at the early drafts, giving suggestions for improvements, and pointing out misattributions and errors. Dr. Catriona Byrne of Springer and her editorial team has my gratitude for being exceptionally helpful and responsive in every step of the preparation of the manuscript.

I would like to acknowledge the role played by the National Science Foundation, the Courant Institute, and UC Berkeley in funding, at various stages, the research related to this book.

And finally, I must thank one person who is outside the realm of mathematics and yet played an indispensable role in the completion of this project: I would have never finished writing the monograph without the sustained urging, care and patience of my wife, Esha. I thank her for all that and much more.

Stanford, CA, USA   Sourav Chatterjee

Contents

1 **Introduction** 1
   1 Superconcentration 1
   2 Chaos 7
   3 Multiple Valleys 11

2 **Markov Semigroups** 15
   1 Semigroup Basics 15
   2 The Ornstein-Uhlenbeck Semigroup 17
   3 Connection with Malliavin Calculus 18
   4 Poincaré Inequalities 18
   5 Some Applications of the Gaussian Poincaré Inequality . . . 19
   6 Fourier Expansion 20

3 **Superconcentration and Chaos** 23
   1 Definition of Superconcentration 23
   2 Superconcentration and Noise-Sensitivity 25
   3 Definition of Chaos 25
   4 Equivalence of Superconcentration and Chaos 27
   5 Some Applications of the Equivalence Theorem 28
   6 Chaotic Nature of the First Eigenvector 29

4 **Multiple Valleys** 33
   1 Chaos Implies Multiple Valleys: The General Idea 33
   2 Multiple Valleys in Gaussian Polymers 34
   3 Multiple Valleys in the SK Model 36
   4 Multiple Peaks in Gaussian Fields 38
   5 Multiple Peaks in the *NK* Fitness Landscape 40

5 **Talagrand's Method for Proving Superconcentration** 45
   1 Hypercontractivity 45
   2 Talagrand's L^1–L^2 Bound 47
   3 Talagrand's Method Always Works for Monotone Functions . . . 49

	4 The Benjamini-Kalai-Schramm Trick	51
	5 Superconcentration in Gaussian Polymers	54
	6 Sharpness of the Logarithmic Improvement	56
6	**The Spectral Method for Proving Superconcentration**	**57**
	1 Spectral Decomposition of the OU Semigroup	58
	2 An Improved Poincaré Inequality	60
	3 Superconcentration in the SK Model	60
7	**Independent Flips**	**63**
	1 The Independent Flips Semigroup	63
	2 Hypercontractivity for Independent Flips	65
	3 Chaos Under Independent Flips	66
8	**Extremal Fields**	**73**
	1 Superconcentration in Extremal Fields	73
	2 A Sufficient Condition for Extremality	78
	3 Application to Spin Glasses	79
	4 Application to the Discrete Gaussian Free Field	83
9	**Further Applications of Hypercontractivity**	**87**
	1 Superconcentration of the Largest Eigenvalue	87
	2 A Different Hypercontractive Tool	90
	3 Superconcentration in Low Correlation Fields	92
	4 Superconcentration in Subfields	93
	5 Discrete Gaussian Free Field on a Torus	94
	6 Gaussian Fields on Euclidean Spaces	99
10	**The Interpolation Method for Proving Chaos**	**105**
	1 A General Theorem	105
	2 Application to the Sherrington-Kirkpatrick Model	111
	3 Sharpness of the Interpolation Method	112
11	**Variance Lower Bounds**	**115**
	1 Some General Tools	115
	2 Application to the Edwards-Anderson Model	118
	3 Chaos in the Edwards-Anderson Model	119
12	**Dimensions of Level Sets**	**125**
	1 Level Sets of Extremal Fields	125
	2 Induced Dimension	128
	3 Dimension of Near-Maximal Sets	131
	4 Applications	134
Appendix A	**Gaussian Random Variables**	**137**
	1 Tail Bounds	137
	2 Size of the Maximum	137
	3 Integration by Parts	139
	4 The Gaussian Concentration Inequality	139
	5 Concentration of the Maximum	140

Appendix B Hypercontractivity	143
References	147
Author Index	153
Subject Index	155

Chapter 1
Introduction

This monograph is about a certain curious feature of random objects that I call 'superconcentration', and two related topics, 'chaos' and 'multiple valleys'. Although superconcentration has been a recognized feature in a number of areas of probability theory in the last twenty years (under a variety of names), its connections with chaos and multiple valleys were discovered and explored for the first time in Chatterjee (2008b). This introductory chapter sketches the basic ideas behind these three concepts through some examples. Precise definitions are given in later chapters.

1 Superconcentration

The theory of concentration of measure gives probability theory a range of tools to compute upper bounds on the orders of fluctuations of complicated random variables. Usually, the general techniques of measure concentration are required when more direct problem-specific approaches do not work. The essential techniques of this theory are adequately summarized in the classic monograph of Ledoux (2001) and the recent book by Boucheron et al. (2013). Roughly speaking, superconcentration happens when the classical measure concentration techniques give suboptimal bounds on the order of fluctuations. The techniques available for proving superconcentration are rather inadequate at this point of time. They give only small improvements on the upper bounds from classical theory. One may wonder what is so interesting about such minor improvements. The main point of this monograph is to demonstrate how *any* improvement over a classical upper bound is equivalent to a number of strange and interesting phenomena, such as chaos under small perturbations and the emergence of multiple valleys.

A formal definition of superconcentration will be given later. Right now, the situation is best explained through examples.

1.1 First-Passage Percolation

Let $\mathbb{Z}^d$ be the integer lattice and let $E(\mathbb{Z}^d)$ be the set of edges of $\mathbb{Z}^d$. Let $(\omega_e)_{e \in E(\mathbb{Z}^d)}$ be a collection of i.i.d. non-negative random variables, called 'edge weights' or 'passage times'. The edge-weight of e models the 'time it takes for a certain fluid to pass through the edge e'. For a path p of connected edges in $\mathbb{Z}^d$, the 'passage time' of p is defined as the sum of passage times of edges along the path. The first-passage time $T(x, y)$ from a vertex x to a vertex y is defined as the minimum over all passage times of paths connecting x to y. Clearly, it suffices to consider self-avoiding paths only. This is the standard model of first-passage percolation, introduced by Hammersley and Welsh (1965) and subsequently studied by numerous authors.

Given $x \in \mathbb{R}^d$ and an integer n, let $T_n(x)$ denote the first-passage time $T(0, [nx])$, where 0 is the origin and $[nx]$ is the lattice point closest to nx. One of the fundamental results of first-passage percolation is that for all x, the limit

$$\lim_{n \to \infty} \frac{T_n(x)}{n}$$

exists and is a deterministic function of x. Moreover, the limit is positive if the probability that an edge weight is zero is less than the critical percolation probability in dimension d. For more on this and other properties of first-passage percolation, see Kesten (1986, 1993) and Grimmett and Kesten (2012).

Therefore under mild conditions (e.g. continuous edge-weights) the first-passage time from 0 to a point at distance n scales like n. For us, the question of interest is: What is the behavior of $\mathrm{Var}(T_n(x))$?

For notational simplicity, let $x = e_1 = (1, 0, 0, \ldots, 0)$, and denote $T_n(x)$ simply by T_n. It was proved by Kesten (1993) using a martingale argument that $\mathrm{Var}(T_n) \leq Cn$, where C is a constant that depends only on the distribution of the edge weights and on the dimension. Kesten moreover proved exponential tail bounds for the fluctuations, which were strengthened by Talagrand (1995) using his famous concentration inequalities for product spaces.

However, none of these bounds improved the bound on the order of fluctuations, which remained at $\sqrt{n}$, until Benjamini et al. (2003) proved that for binary edge weights in dimension $d \geq 2$,

$$\mathrm{Var}(T_n) \leq \frac{Cn}{\log n}.$$

Here again, C is a constant depending only on the distribution of the edge-weights and the dimension. This result was extended to a more general class of edge weights by Benaïm and Rossignol (2008).

The Benjamini-Kalai-Schramm (BKS) theorem is an example of what I call 'superconcentration', which means, roughly, that the order of fluctuations is less than the upper bound given by classical theory. A more precise definition will be given later.

The physicists conjecture that when $d = 2$, $\text{Var}(T_n)$ should scale like $n^{2/3}$. Besides numerical evidence, there is also some indirect mathematical evidence to support this conjecture: The closely related model of two-dimensional oriented last-passage percolation is 'exactly solvable' in the case of exponential or geometric edge-weights, famously proved by Johansson (2000), who derived the $n^{1/3}$ order of fluctuations as a consequence of the exact solution. See for example Borodin et al. (2012) for more details and a survey of exciting recent developments.

In higher dimensions, there is no consensus about the behavior of the variance of T_n. Some experts think that $\text{Var}(T_n)$ remains bounded as $n \to \infty$ in sufficiently high dimension, while others disagree.

It is not hard to prove that $\text{Var}(T_n)$ is bounded below by a positive constant depending on edge weights and the dimension. A non-trivial lower bound was proved by Newman and Piza (1995) and simultaneously by Pemantle and Peres (1994), who showed that in $d = 2$, $\text{Var}(T_n)$ must grow at least as fast as a multiple of $\log n$ under mild conditions.

In Chap. 5, we will see a proof of the following variant of the BKS theorem.

Theorem 1.1 (Variant of Theorem 1 in Benjamini et al. 2003) *Consider first-passage percolation on $\mathbb{Z}^d$, $d \geq 2$. If the edge-weight distribution can be realized as the probability distribution of a non-negative Lipschitz function of a Gaussian random variable that is uniformly bounded away from zero and infinity, then for all n,*

$$\text{Var}(T_n) \leq \frac{Cn}{\log n},$$

where C is a constant that depends only on the distribution of the edge-weights and the dimension.

The class of edge-weight distributions covered by the above theorem includes, for instance, uniform distributions on bounded intervals.

Open Problem 1.2 Improve the upper bound on the variance in Theorem 1.1. Preferably, find the correct order of the variance.

1.2 Gaussian Random Polymers

Let n be a positive integer, and consider the set of all 1-dimensional random walk paths of length n, starting at 0. In other words, each path is a sequence like $\{(0, a_0), (1, a_1), \ldots, (n, a_n)\}$, where $a_0 = 0$, and $|a_{i+1} - a_i| = 1$ for each i, so that there are 2^n such paths. Any such path is a possible shape of a $(1 + 1)$-dimensional polymer. Here $(1 + 1)$ means 1 space dimension and 1 time dimension.

Let $(g_v)_{v \in \mathbb{Z}^2}$ be a collection of i.i.d. standard Gaussian random variables, called the 'environment' or the 'medium'. Given a random walk path p of length n, define the 'energy' of p as

$$H_n(p) := -\sum_{v \in p} g_v.$$

This is the so-called 'Gaussian random polymer'—or more correctly, the '(1 + 1)-dimensional polymer in Gaussian random media'—originally defined in the mathematical physics literature (see Imbrie and Spencer 1988). The model generalizes easily in higher dimensions, leading to the so-called $(d + 1)$-dimensional polymer.

An object of interest is the minimum energy of a path of length n. This is called the 'ground state energy', which will be denoted by E_n. Another object of interest is the minimum energy path, which will be denoted by $\hat{p}_n$. Incidentally, the $(1 + 1)$-dimensional random polymer model is exactly the same as the model of two-dimensional oriented last-passage percolation.

Along the lines of the BKS theorem (but with some additional technical difficulties), the following result was proved in Chatterjee (2008b).

Theorem 1.3 (Theorem 8.1 in Chatterjee 2008b) *If E_n is the ground state energy in the $(1 + 1)$-dimensional Gaussian random polymer model, then*

$$\mathrm{Var}(E_n) \le \frac{Cn}{\log n},$$

where C does not depend on n.

We will see a proof of this theorem in Chap. 5. The result was extended to $(d+1)$-dimensional polymers by Graham (2010).

Open Problem 1.4 Improve the upper bound on the variance in Theorem 1.3. Preferably, find the correct order of the variance.

As in first-passage percolation, classical techniques give a bound of order n instead of $n/\log n$. Theorem 1.3 is another instance of superconcentration.

Incidentally, the $\log n$ improvement over the classical bounds may be optimal in certain examples, although not in first-passage percolation or directed polymers. This is demonstrated by the following example. Let $(g_v)_{v \in \mathbb{Z}^2}$ be i.i.d. standard Gaussian random variables. Consider all 'paths' of the form

$$\{(1, a_1), (2, a_2), \ldots, (n, a_n)\},$$

where now the a_i's can take any value in $\{1, \ldots, n\}$, with no restrictions. Then clearly the maximum possible value of the sum of vertex weights along such a path is precisely

$$\sum_{i=1}^{n} \max_{1 \le j \le n} g_{(i,j)},$$

which has variance of order $n/\log n$ since the maximum of n i.i.d. standard Gaussian random variables has variance of order $1/\log n$ (see Appendix A).

1.3 The Sherrington-Kirkpatrick Model of Spin Glasses

In this model there are n particles, each carrying a spin of $+1$ or -1. A spin configuration $\sigma = (\sigma_1, \ldots, \sigma_n)$ is an element of $\{-1, 1\}^n$. Let $(g_{ij})_{1 \leq i < j \leq n}$ be a collection of i.i.d. standard Gaussian random variables, called the 'disorder'. Given the disorder, define the energy of a spin configuration σ as

$$H_n(\sigma) = -\frac{1}{\sqrt{n}} \sum_{1 \leq i < j \leq n} g_{ij} \sigma_i \sigma_j.$$

This defines the so-called Sherrington-Kirkpatrick (SK) model of spin glasses introduced by Sherrington and Kirkpatrick (1975).

Having so defined the energy, the 'free energy' at 'inverse temperature' $\beta \geq 0$ is defined as

$$F_n(\beta) = -\frac{1}{\beta} \log Z_n(\beta),$$

where

$$Z_n(\beta) = \sum_{\sigma \in \{-1,1\}^n} e^{-\beta H_n(\sigma)}.$$

It is known that

$$\lim_{n \to \infty} \frac{F_n(\beta)}{n}$$

exists and is a deterministic function of β. The existence of the limit was proved using an ingenious subadditive argument by Guerra and Toninelli (2002). The expression for the limit is given by the remarkable 'Parisi formula', proved rigorously in a famous paper of Talagrand (2006), who built upon an interpolation scheme introduced by Guerra (2003). Extensive details about this and other models of spin glasses, as well as references to the literature, are available in Talagrand (2011, 2012) and Panchenko (2013a,b).

We are interested in the order of fluctuations of $F_n(\beta)$. When $\beta < 1$, it was proved by Aizenman et al. (1987) that

$$\lim_{n \to \infty} \mathrm{Var}(F_n(\beta))$$

exists and is finite. They moreover gave a formula for this limit and proved a central limit theorem for $F_n(\beta)$. Possibly, this result can be extended to $\beta = 1$ to prove an optimal bound for the variance.

Open Problem 1.5 Analyze the fluctuations of $F_n(\beta)$ when $\beta = 1$.

When $\beta > 1$, the best known upper bound for a long time was the classical bound $\mathrm{Var}(F_n(\beta)) \leq Cn$. (See Talagrand 2003, Corollary 2.2.5, for example.) Superconcentration of the free energy was proved for the first time in Chatterjee (2009).

Theorem 1.6 (Theorem 1.5 in Chatterjee 2009) *If $F_n(\beta)$ is the free energy of the SK model at inverse temperature β, then*

$$\operatorname{Var}(F_n(\beta)) \leq \frac{C(\beta)n}{\log n},$$

where $C(\beta)$ depends only on β. This result holds for all $\beta \geq 0$.

Interestingly, though the result looks similar to the BKS theorem, the method of proof is very different. The hypercontractive tools that lie at the heart of the BKS proof do not seem to work for spin glasses.

I have heard it said that $\operatorname{Var}(F_n(\beta))$ should be $O(1)$ for all β, or at least for $\beta \neq 1$.

Open Problem 1.7 Find the correct order of $\operatorname{Var}(F_n(\beta))$. In particular, try to prove that it is $O(1)$ for all β.

See Parisi and Rizzo (2009) for some physics conjectures and predictions related to the above problem.

In Chap. 10, we will see a proof of Theorem 1.6. A slightly worse bound with an easier proof will be presented in Chap. 6.

1.4 Maxima of Gaussian Fields

Consider a Gaussian field $g = (g_1, \ldots, g_n)$ on the finite index set $\{1, \ldots, n\}$. This is just an n-dimensional Gaussian random vector, but I use the term 'field' because there is a geometry involved, namely the geometry induced by the L^2 metric

$$d(i, j) := \left(\mathbb{E}(g_i - g_j)^2\right)^{1/2}.$$

Assume that g is centered, i.e. each g_i has mean zero. Let

$$R(i, j) := \operatorname{Cov}(g_i, g_j).$$

The question of interest is the following: What is the variance of $\max_i g_i$? It is a well-known result that

$$\operatorname{Var}\left(\max_{1 \leq i \leq n} g_i\right) \leq \max_{1 \leq i \leq n} \operatorname{Var}(g_i). \tag{1.1}$$

This inequality was proved by Houdré (1995), although the method of proof seems to be implicit in the much earlier work of Nash (1958), and the works of Chernoff (1981), Chen (1982), and Houdré and Kagan (1995). In Chap. 2, we will see how to prove this inequality.

The inequality (1.1) is easily seen to be tight by taking $R(i, j) = 1$ for all i, j. However it can often be suboptimal, e.g. when the g_i's are independent. In fact, the ground state energy of Gaussian random polymers is a special case of this general problem. When we formally define superconcentration in Chap. 3, we will see that the maximum of a Gaussian field is superconcentrated if and only if inequality (1.1) is suboptimal.

Open Problem 1.8 Give a general and easily verifiable condition on the covariance matrix R that is equivalent to superconcentration of the maximum.

2 Chaos

Traditionally, chaos is defined as 'high sensitivity to initial conditions' in the context of dynamical systems. In the statistical physics of disordered systems (such as spin glasses), chaos takes on a slightly different but related meaning: a system in statistical physics is called chaotic if it is highly sensitive to small changes in parameters such as temperature, disorder, etc. While this is not a precise definition, its meaning may be easily clarified through examples.

2.1 Chaos in Gaussian Polymers

While studying random polymer models, physicists (e.g. Huse et al. 1985, Zhang 1987, Mézard 1990) have studied whether polymers are chaotic, that is, whether they are sensitive to small changes in the environment. A specific object of interest is the polymer with minimum energy, i.e. the ground state polymer. Recall that we introduced this model in Sect. 1.2, and denoted the ground state by $\hat{p}_n$.

The physicists have a standard way of perturbing a Gaussian environment. Consider the $(1 + 1)$-dimensional Gaussian polymer, for instance. Take a $t > 0$ and for each vertex $v \in \mathbb{Z}^2$, let (g_v, g_v^t) be a pair of jointly Gaussian random variables with mean 0, variance 1 and correlation e^{-t}. When t is close to zero, $g_v \approx g_v^t$ with high probability. Let $\{(g_v, g_v^t) : v \in \mathbb{Z}^2\}$ be a collection of i.i.d. pairs. Then if $(g_v)_{v \in \mathbb{Z}^2}$ is our original environment, we may call $(g_v^t)_{v \in \mathbb{Z}^2}$ the perturbed environment, or more precisely, the t-perturbed environment.

The following theorem was proved in Chatterjee (2008b). It may be proved using a consequence of Theorem 1.6 from Sect. 1.3 of this chapter and Theorem 3.5 from Chap. 3.

Theorem 1.9 (Theorem 8.1 in Chatterjee 2008b) *Consider the $(1+1)$-dimensional Gaussian random polymer of length n. Let $\hat{p}_n$ denote the ground state polymer in the original environment, and let $\hat{p}_n^t$ denote the ground state polymer in the t-perturbed* environment. Let $|\hat{p}_n \cap \hat{p}_n^t|$ denote the number of vertices in the intersection of the two polymer paths. Then

$$\mathbb{E}|\hat{p}_n \cap \hat{p}_n^t| \le \frac{Cn}{(1-e^{-t})\log n},$$

where C is a constant that does not depend on n or t.

The above result shows that the paths are 'almost disjoint' relative to their lengths when $t \gg 1/\log n$. This result is related to the formula

$$\text{Var}(E_n) = \int_0^\infty e^{-t} \mathbb{E}|\hat{p}_n \cap \hat{p}_n^t| \, dt, \qquad (1.2)$$

first proved in Chatterjee (2008b). This is a special case of the so-called 'dynamical formulas of variance' that are widely used in the literature on Gaussian fields (see e.g. Deuschel et al. 2000, Proposition 2.2 or Adler and Taylor 2007, Chap. 2). We will see a proof of this identity and its relation to Theorem 1.9 later in the monograph.

For the physics connection to Theorem 1.9, see for example da Silveira and Bouchaud (2004).

One of the main thrusts of this monograph is that a superconcentration result such that Theorem 1.3 and a chaos result such as Theorem 1.9 are essentially equivalent. Indeed, we will see how to prove the following theorem in Chap. 3.

Theorem 1.10 *Consider the Gaussian random polymer model in any dimension. If E_n is the ground state energy, then*

$$\text{Var}(E_n) = o(n)$$

if and only if there exists $t_n \to 0$ such that

$$\mathbb{E}|\hat{p}_n \cap \hat{p}_n^{t_n}| = o(n).$$

Combining this with the superconcentration result of Graham (2010), it follows that the Gaussian random polymer is chaotic in any dimension.

2.2 Chaos in the SK Model

Given an inverse temperature β and a disorder $(g_{ij})_{1 \le i < j \le n}$, the SK model defines a 'Gibbs measure' on $\{-1, 1\}^n$ by putting mass

$$Z_n(\beta)^{-1} e^{-\beta H_n(\sigma)}$$

at each configuration σ, where Z_n and H_n are defined in Sect. 1.3. The Gibbs measure is a random probability measure on $\{-1, 1\}^n$.

2 Chaos

An important quantity associated with the SK model is the 'overlap': Having defined the Gibbs measure, let σ^1 and σ^2 be two spin configurations drawn independently from the Gibbs measure. The overlap between σ^1 and σ^2 is defined as

$$R_{1,2} := \frac{1}{n} \sum_{i=1}^{n} \sigma_i^1 \sigma_i^2. \tag{1.3}$$

It is known from the work of Aizenman et al. (1987) that when $\beta < 1$,

$$\lim_{n \to \infty} \mathbb{E}(R_{1,2}^2) = 0.$$

It was shown later by Guerra (1995) that the same result holds when $\beta = 1$. However, it is believed that the system undergoes a phase transition at $\beta = 1$, one of the effects of which is that the above limit is positive when $\beta > 1$. (For a proof of this, see Example 1 in Panchenko 2008.)

The chaos question in the SK model is similar to the one in polymers: Is the Gibbs measure chaotic under small perturbations of the disorder? (This is known as chaos in disorder; there is also a similar 'chaos in temperature'.)

The question is precisely formulated as follows. As in the polymer model, let $(g_{ij}^t)_{1 \le i < j \le n}$ be a t-perturbed disorder; this means, for each (i, j), (g_{ij}, g_{ij}^t) is a pair of Gaussian random variables with mean zero, variance 1, and correlation e^{-t}.

Fix $\beta \ge 0$. Let σ^1 be a configuration picked from the original Gibbs measure at inverse temperature β, and let σ^2 be a configuration picked from the Gibbs measure defined by the t-perturbed disorder. Given the two disorders, σ^1 and σ^2 are independent. Define

$$R_{1,2}(t) = \frac{1}{n} \sum_{i=1}^{n} \sigma_i^1 \sigma_i^2.$$

Physicists (e.g. Bray and Moore 1987, Fisher and Huse 1986, Krząkała and Bouchaud 2005) define chaos in disorder to be the phenomenon that $R_{1,2}(t)$ is close to zero for some small positive t. (For an up-to-date survey of the physics literature on this, see Rizzo 2009.) This is easily proved to be true when $\beta < 1$ following the Aizenman-Lebowitz-Ruelle argument; the point is that this is supposed to be true for all β. This was proved rigorously in Chatterjee (2009).

Theorem 1.11 (Theorem 1.3 in Chatterjee 2009) *Let $R_{1,2}(t)$ be defined as above. Then for any integer $k \ge 1$,*

$$\mathbb{E}(R_{1,2}^{2k}(t)) \le (C_1(\beta)k)^k n^{-C_2(\beta)k \min\{1,t\}},$$

where $C_1(\beta)$ and $C_2(\beta)$ are positive constants that depend only on β.

A non-trivial extension of the above theorem to the SK model in the presence of an external field was recently achieved by Chen (2011). Chaos in mixed p-spin models was proved by Chen and Panchenko (2012).

The above theorem will be proved in Chap. 10. Moreover, we will show in Chap. 3 that this theorem is equivalent to the superconcentration of the free energy:

Theorem 1.12 *In the SK model,*

$$\mathrm{Var}(F_n(\beta)) = o(n)$$

if and only if there exists $t_n \to 0$ such that

$$\mathbb{E}(R_{1,2}^2(t_n)) = o(1).$$

2.3 Chaos in Gaussian Fields

Let $g = (g_1, \ldots, g_n)$ be a centered Gaussian field with covariance matrix R, as in Sect. 1.4. Let g' be an independent copy of g. For each $t \geq 0$, let

$$g^t := e^{-t} g + \sqrt{1 - e^{-2t}} g'.$$

Under the mild condition that $R(i, j) \neq 1$ for all $i \neq j$, there is almost surely a unique (random) index $I \in \{1, \ldots, n\}$ where the field g attains its maximum. Similarly, for each $t \geq 0$ let I^t be the unique index at which g^t attains its maximum. Note that $g^0 = g$ and $I^0 = I$.

Suppose that the field g is 'positively correlated', i.e. $R(i, j) \geq 0$ for all i, j. Note that in the L^2 geometry on $\{1, \ldots, n\}$ induced by the field g, the distance $d(i, j)$ may be represented as

$$d(i, j) = \sqrt{R(i, i) + R(j, j) - 2R(i, j)}.$$

In other words, the smaller the value of $R(i, j)$, the further apart are the points i and j. Thus one may roughly say that the location of the maximum of g is 'chaotic under small perturbations' if $\mathbb{E}(R(I^0, I^t))$ is close to zero for all values of t above a small threshold.

The following generalization of the identity (1.2) was first proved in Chatterjee (2008b) and simultaneously in Nourdin and Viens (2009):

$$\mathrm{Var}\left(\max_{1 \leq i \leq n} g_i\right) = \int_0^\infty e^{-t} \mathbb{E}(R(I^0, I^t)) \, dt. \tag{1.4}$$

The important observation made in Chatterjee (2008b) is that the integrand on the right is always a decreasing function of t. From this the following theorem was derived. It shows that the maximum of a Gaussian field is superconcentrated if and only if the location of the maximum is chaotic under small perturbations. We will see how to prove this theorem in Chap. 3.

Theorem 1.13 (Theorem 3.2 in Chatterjee 2008b) *Let g, R and I^t be as above. Let $v = \text{Var}(\max_{1 \leq i \leq n} g_i)$ and $w := \max_{1 \leq i \leq n} \text{Var}(g_i)$. Then for each $t \geq 0$ we have*

$$0 \leq \mathbb{E}(R(I^0, I^t)) \leq \frac{v}{1 - e^{-t}}, \quad \text{and}$$

$$v \leq w(1 - e^{-t}) + \mathbb{E}(R(I^0, I^t))e^{-t}.$$

To understand this result, suppose that the coordinates are normalized to satisfy $\text{Var}(g_i) = 1$ for each i. Then $w = 1$, and the maximum is superconcentrated if and only if v is close to zero. If $v \approx 0$, then the first inequality shows that $\mathbb{E}(R(I^0, I^t)) \approx 0$ for all t above a small threshold, namely, for $t \gg v$. On the other hand, if $\mathbb{E}(R(I^0, I^t)) \approx 0$ for all t above a small threshold, let's say $t \geq \delta$, then the second inequality shows that $v \ll w$, e.g. by taking $t = \delta$. Thus, the first inequality proves that superconcentration of the maximum implies that the location of the maximum is chaotic under small perturbations, and the second inequality proves the converse statement.

Note that the condition that $R(i, j) \geq 0$ for all i, j is not required for the above theorem to hold, but the result does not carry a lot of meaning without this condition.

3 Multiple Valleys

This section is an introduction to the third property related to superconcentration, namely, multiple valleys. To motivate the definition of multiple valleys, we will first discuss a related concept that is in some sense the opposite of multiple valleys.

3.1 Asymptotic Essential Uniqueness

An optimization problem is called 'stable' if any near-optimum is close to the optimum in some appropriate metric. Mathematics abounds with stability questions for optimization problems. In probability theory, we usually have a sequence of optimization problems instead of a single problem. Aldous (2001) defined a notion of stability that makes sense in the probabilistic setting. This is known as Asymptotic Essential Uniqueness (AEU).

The first example where AEU was rigorously established is the so-called random assignment problem (see Aldous 2001). There are n tasks and n individuals. Each individual can be assigned exactly one task. The cost of assigning task j to individual i is c_{ij}. The assignment problem seeks to minimize $\sum_{i=1}^{n} c_{i\pi(i)}$ over all permutations π of $\{1, \ldots, n\}$. When the c_{ij}'s are random variables, probability enters into the picture. In the simplest scenario, the c_{ij}'s are i.i.d. non-negative random variables. Assume that the c_{ij}'s have a density with respect to Lebesgue measure and that the value of the density at 0 is 1. (For example, the c_{ij}'s may be i.i.d. exponential random variables with mean 1.) Aldous (2001) proved a famous conjecture of

Parisi, who claimed that as $n \to \infty$, the minimum cost of the random assignment problem approaches $\zeta(2)$, where ζ is the Riemann zeta function.

In the same paper, Aldous introduced and proved the AEU property for the random assignment problem. It was shown that any assignment (permutation) that nearly minimizes the cost must be almost the same as the optimal assignment. The precise result is as follows: For all $0 < \delta < 1$, there exists an $\epsilon(\delta) > 0$ such that if μ_n are (random) permutations depending on $(c_{ij})_{1 \le i, j \le n}$ such that for all n,

$$\mathbb{E}\left(\frac{1}{n}|\{i : \mu_n(i) \ne \pi_n(i)\}|\right) \ge \delta,$$

then

$$\liminf_{n \to \infty} \mathbb{E}\left(\sum_{i=1}^{n} c_{i\mu(i)}\right) \ge \zeta(2) + \epsilon.$$

Similar results for the minimal spanning tree on graphs with randomly weighted edges and a few other models were later proved by Aldous et al. (2008, 2009). Other problems, such as the random Euclidean traveling salesman problem, are thought to have the AEU property but with no proofs (see Aldous 2001, Sect. 7).

A simple example where AEU can be directly verified is as follows. Let $g_1, \ldots, g_n$ be i.i.d. standard Gaussian random variables. Define a random function $f_n : \{-1, 1\}^n \to \mathbb{R}$ as

$$f_n(\sigma) = \sum_{i=1}^{n} g_i \sigma_i.$$

The problem is to maximize this function over all possible values of σ. Clearly, the maximum is attained at $\hat{\sigma}$, where $\hat{\sigma}_i = \text{sign}(g_i)$. It is simple to prove that this optimization problem has the AEU property. More precisely, one can try to prove the following exercise.

Exercise 1.14 Show that for any given $\epsilon > 0$ there is a $\delta > 0$ small enough, such that with probability tending to 1 as $n \to \infty$, all σ such that $f_n(\sigma) \ge (1 - \delta) f_n(\hat{\sigma})$ satisfy

$$\frac{1}{n}|\{i : \sigma_i \ne \hat{\sigma}_i\}| \le \epsilon.$$

3.2 Multiple Valleys and Peaks

The notion of multiple valleys (or multiple peaks) is sort of an opposite of the AEU property. This concept was introduced in Chatterjee (2008b) and is somewhat different than the notion of multiple valleys in the physics folklore.

3 Multiple Valleys

Roughly speaking, we will say that a random optimization problem has multiple valleys if there are many vastly dissimilar near-optimal solutions. A precise definition may be given as follows. Suppose we have a sequence of sets X_n, and for each n, $f_n : X_n \to \mathbb{R}$ is a random function. Let s_n be a 'similarity measure' on X_n: for each $x, y \in X_n$, $s_n(x, y)$ is a non-negative real number that measures the degree of similarity between x and y. We impose no conditions on s_n besides non-negativity (and measurability, if the situation demands).

Definition 1.15 A sequence (f_n, X_n, s_n) (or simply f_n) is said to have the Multiple Valley (MV) property if there exist ϵ_n, δ_n and γ_n tending to zero and K_n tending to infinity, such that for each n, with probability $\geq 1 - \gamma_n$, there exists a set $A \subseteq X_n$ of cardinality $\geq K_n$ such that $s_n(x, y) \leq \epsilon_n$ for all $x, y \in A$, $x \neq y$, and for all $x \in A$,

$$\left| \frac{f_n(x)}{\min_{y \in X_n} f_n(y)} - 1 \right| \leq \delta_n.$$

If the min is replaced by max, we shall call it the Multiple Peaks property.

Let us now go back to the Sherrington-Kirkpatrick model. Take X_n to be the space $\{-1, 1\}^n$, f_n to be the Hamiltonian function

$$H_n(\sigma) = -\frac{1}{\sqrt{n}} \sum_{1 \leq i < j \leq n} g_{ij} \sigma_i \sigma_j,$$

and s_n to be the similarity measure

$$s_n(\sigma, \sigma') = \left(\frac{1}{n} \sum_{i=1}^{n} \sigma_i \sigma_i' \right)^2.$$

The following result was first proved in Chatterjee (2009).

Theorem 1.16 (Chatterjee 2009) *For the similarity measure defined above, the Hamiltonian of the Sherrington-Kirkpatrick model has the multiple valley property. In other words, when n is large, then with high probability there are many states with near-minimal energy that are all nearly mutually orthogonal to each other.*

The connection with our previous discussion of chaos and superconcentration is that we will derive the multiple valley result as a consequence of the chaos property of the SK model.

We will prove a similar result for $(1 + 1)$-dimensional polymers. Here the space X_n is the space of all random walk paths of length n. The similarity between two paths p and p' is

$$s_n(p, p') = \frac{|p \cap p'|}{n},$$

where $|p \cap p'|$ is the number of vertices in the intersection of the two paths. The function f_n in this case is the Hamiltonian (energy) function H_n, that is, the sum of vertex weights along a given path, together with a negative sign in front. The following theorem will be proved as a consequence of the superconcentration of the ground state energy. This was originally proved in Chatterjee (2008b).

Theorem 1.17 (Chatterjee 2008b) *The energy function in the $(1+1)$-dimensional Gaussian polymer model has multiple valleys, if the similarity measure is defined as above. In other words, when n is large, then with high probability there are many paths that all have nearly minimal energy and are all nearly disjoint from each other.*

The result extends to the $(d+1)$-dimensional polymer for any $d \geq 2$ using the superconcentration result of Graham (2010).

In Chap. 4, we will see how the existence of multiple valleys follows from superconcentration. Combining this with the superconcentration of the free energy (in the case of the SK model) and the ground state energy (in the case of polymers), the proofs of Theorems 1.17 and 1.16 will follow.

For general Gaussian fields, it was proved in Chatterjee (2008b) that superconcentration (or equivalently, chaos) implies the existence of multiple peaks. The following theorem is a cleaned-up version of the result from Chatterjee (2008b).

Theorem 1.18 (Variant of Theorem 3.7 in Chatterjee 2008b) *Let $g = (g_1, \ldots, g_n)$ be a centered Gaussian field with covariance matrix R. Suppose that $R(i,i) = 1$ and $R(i,j) \geq 0$ for all i,j, and let*

$$\epsilon := \mathrm{Var}\Big(\max_{1 \leq i \leq n} g_i\Big).$$

Then there is a universal constant C such that if

$$\delta := \frac{C}{\sqrt{\log(1/\epsilon)}},$$

then with probability at least $1 - \delta$, there are at least $1/\delta$ points i satisfying

$$g_i \geq (1 - \delta) \max_{1 \leq j \leq n} g_j,$$

such that for any two of these points i and j, $R(i,j) \leq \delta$.

In other words, if ϵ is close to zero (i.e. the maximum is superconcentrated), then with high probability there are a large number of indices that are mutually 'distant' from each other, at which g is near-maximal.

Chapter 2
Markov Semigroups

This chapter introduces a key tool in the analysis of measure concentration, namely, the tool of semigroup analysis. After introducing the basics of Markov semigroups, we will see how the semigroup tools may be applied to prove Poincaré inequalities. Poincaré inequalities, which give variance bounds, are the most elementary among all concentration inequalities. We will then see how to apply Poincaré inequalities to get variance bounds in a number of non-trivial problems. The results of this chapter, in various guises, may be found in standard sources on concentration of measure, e.g. the monograph of Ledoux (2001), the lecture notes of Guionnet and Zegarlinski (2003), etc. Since the results are well known, we will only present the essential components and omit technical details.

1 Semigroup Basics

Suppose that $(X_t)_{t\geq 0}$ is a Markov process taking values in some abstract state space. The Markov process naturally defines a semigroup of operators $(P_t)_{t\geq 0}$ acting on functions from the state space into the real line:

$$P_t f(x) = \mathbb{E}\big(f(X_t) \,\big|\, X_0 = x\big),$$

assuming, of course, that the right-hand side is well-defined. This is a semigroup since $P_{t+s} = P_t P_s$. The generator L of the semigroup is the operator

$$Lf = \lim_{t \to 0} \frac{P_t f - f}{t} = \partial_t P_t \bigg|_{t=0},$$

again assuming that the right-hand side is well-defined for the given f. The *heat equation* for the semigroup P_t is the equation

$$\partial_t P_t = L P_t.$$

This is a simple consequence of the definition of L and the semigroup nature of P_t. It follows easily from the heat equation that

$$P_t = e^{tL} = \sum_{k=0}^{\infty} \frac{t^k}{k!} L^k.$$

Usually L is bounded in some norm, so that the series converges.

Assume that the Markov process X_t has an equilibrium probability measure μ. This means that no matter which state it starts out from, the limiting distribution of X_t as $t \to \infty$ is μ. The measure μ automatically defines an L^2 space and an inner product:

$$(f, g) := \int fg \, d\mu.$$

Moreover, it also defines an important bilinear form $\mathcal{E}$ called the Dirichlet form of the Markov semigroup P_t:

$$\mathcal{E}(f, g) := -(f, Lg) = -\int f Lg \, d\mu.$$

When L is self-adjoint, the Dirichlet form is symmetric. This happens when the Markov process is reversible. From now on, we will deal with reversible processes only.

Another important bilinear form is the covariance:

$$\mathrm{Cov}_\mu(f, g) := \int fg \, d\mu - \int f \, d\mu \int g \, d\mu.$$

The variance is defined as $\mathrm{Var}_\mu(f) = \mathrm{Cov}_\mu(f, f)$. In the same vein, we will often write $\int f \, d\mu$ as $\mathbb{E}_\mu(f)$.

The following simple but important result will be called 'the covariance lemma' in this monograph.

Lemma 2.1 *For any $f, g \in L^2(\mu)$,*

$$\mathrm{Cov}_\mu(f, g) = \int_0^\infty \mathcal{E}(f, P_t g) \, dt,$$

provided that the derivative can be moved inside the integral when differentiating $(f, P_t g)$ with respect to t, and the heat equation $\partial_t P_t g = L P_t g$ holds.

Proof It is implicit in the assumptions that $(f, P_t g)$ is differentiable in t. Recall that $P_t g$ tends to $\mathbb{E}_\mu(g)$ in L^2 as $t \to \infty$. Therefore, making use of the assumptions of the lemma,

$$\mathrm{Cov}_\mu(f, g) = (f, g) - \lim_{t \to \infty} (f, P_t g)$$

$$= -\int_0^\infty \partial_t(f, P_t g)dt$$
$$= -\int_0^\infty (f, \partial_t P_t g)dt$$
$$= -\int_0^\infty (f, L P_t g)dt = \int_0^\infty \mathcal{E}(f, P_t g)dt.$$

This completes the proof of the lemma. □

The two technical assumptions of Lemma 2.1 will be satisfied in all our examples.

2 The Ornstein-Uhlenbeck Semigroup

The standard Ornstein-Uhlenbeck (OU) process is a process $(X_t)_{t \geq 0}$ taking values in $\mathbb{R}$ that satisfies the stochastic differential equation

$$dX_t = -X_t\, dt + \sqrt{2}\, dB_t,$$

where $(B_t)_{t \geq 0}$ is a standard Brownian motion. This is a continuous version of the discrete time autoregressive process. An alternative way to represent the OU process is to write X_t as

$$X_t = e^{-t} X_0 + e^{-t} W_{e^{2t}-1},$$

where $(W_s)_{s \geq 0}$ is again a standard Brownian motion (although not the same as B_t) and X_0 is the initial state (that is independent of the W process). The alternative representation shows that the semigroup corresponding to the OU process has the formula

$$P_t f(x) = \mathbb{E}\big(f\big(e^{-t}x + \sqrt{1 - e^{-2t}}\, Z\big)\big),$$

where Z is a standard Gaussian random variable. The formula clearly shows that the standard Gaussian distribution is the equilibrium measure for the OU process.

Differentiation with respect to t and using the Gaussian integration by parts identity $\mathbb{E}(Zg(Z)) = \mathbb{E}(g'(Z))$ shows quite easily that the generator of the process is

$$Lf(x) = \partial_t P_t f(x)|_{t=0} = f''(x) - xf'(x),$$

provided that f is twice differentiable and derivatives do not grow too fast at infinity.

Similarly, a further application of integration by parts shows that the Dirichlet form is

$$\mathcal{E}(f, g) = \mathbb{E}\big(f'(Z)g'(Z)\big).$$

The OU process has a simple n-dimensional version: Each coordinate is a one-dimensional OU process, and the processes are independent. Simple verifications as above show that for this process,

$$P_t f(x) = \mathbb{E}\big(f\big(e^{-t}x + \sqrt{1-e^{-2t}}Z\big)\big) \tag{2.1}$$

where Z is an n-dimensional standard Gaussian random vector, and

$$Lf(x) = \Delta f(x) - x \cdot \nabla f(x),$$

where Δ is the Laplacian operator and ∇ is the gradient. Similarly,

$$\mathcal{E}(f, g) = \mathbb{E}_{\gamma^n}(\nabla f \cdot \nabla g), \tag{2.2}$$

where γ^n is the n-dimensional standard Gaussian measure.

3 Connection with Malliavin Calculus

For the Ornstein-Uhlenbeck semigroup, the covariance lemma (Lemma 2.1) may be obtained as a consequence of the so-called 'Mehler formula' from Malliavin calculus (see e.g. Nualart 2009 for the statement of Mehler's formula). This observation was made in Nourdin and Viens (2009) and Nourdin and Peccati (2009), and subsequently exploited in the recent body of work connecting Stein's method of normal approximation with Malliavin calculus. In particular, the formula (1.4) for the variance of the maximum of a Gaussian field was derived using Mehler's formula in Nourdin and Viens (2009) around the same time that it appeared in Chatterjee (2008b), and was used in the context of estimating bounds on probability densities of Gaussian functionals. The nice monograph of Nourdin and Peccati (2012) gives a comprehensive survey of this emerging area. We will not delve further into this, because the connection between Malliavin calculus and superconcentration seems to go only as far as the covariance lemma and the formula (1.4).

4 Poincaré Inequalities

Suppose that we have a Markov semigroup P_t with Dirichlet form $\mathcal{E}$ and equilibrium measure μ. This semigroup is said to satisfy a Poincaré inequality with constant C if for all $f \in L^2(\mu)$,

$$\mathrm{Var}_\mu(f) \leq C\,\mathcal{E}(f, f)$$

whenever the right-hand side is well-defined. The smallest C for which this is satisfied is called the 'optimal constant' in the Poincaré inequality.

Below, we show that the n-dimensional Ornstein-Uhlenbeck semigroup satisfies a Poincaré inequality with optimal constant 1.

The representation (2.1) of the n-dimensional OU semigroup shows that

$$\nabla P_t f = e^{-t} P_t \nabla f, \tag{2.3}$$

where $P_t \nabla f$ is the vector whose ith coordinate is $P_t \partial_i f$, where $\partial_i f$ is the partial derivative of f in the ith coordinate.

Therefore by the covariance lemma (Lemma 2.1),

$$\begin{aligned}
\operatorname{Var}_{\gamma^n}(f) &= \int_0^\infty \mathcal{E}(f, P_t f)\, dt \\
&= \int_0^\infty \mathbb{E}_{\gamma^n}(\nabla f \cdot \nabla P_t f)\, dt \\
&= \int_0^\infty e^{-t} \mathbb{E}_{\gamma^n}(\nabla f \cdot P_t \nabla f)\, dt.
\end{aligned} \tag{2.4}$$

By the Cauchy-Schwarz inequality,

$$\nabla f \cdot P_t \nabla f \leq |\nabla f| |P_t \nabla f|,$$

where $|\cdot|$ denotes the Euclidean norm. Again by Cauchy-Schwarz,

$$\mathbb{E}_{\gamma^n} |\nabla f| |P_t \nabla f| \leq \left(\mathbb{E}_{\gamma^n} |\nabla f|^2\, \mathbb{E}_{\gamma^n} |P_t \nabla f|^2 \right)^{1/2}.$$

By Jensen's inequality, $\mathbb{E}((P_t h)^2) \leq \mathbb{E}(h^2)$ for any real valued function h. Combining the last four steps we get

$$\operatorname{Var}_{\gamma^n}(f) \leq \mathbb{E}_{\gamma^n} |\nabla f|^2 = \mathcal{E}(f, f). \tag{2.5}$$

This is known as the Gaussian Poincaré inequality. It is easy to fill in the technical details to prove that this inequality holds for any absolutely continuous f such that the right-hand side is finite.

Inequality (2.5) shows that the n-dimensional OU semigroup satisfies a Poincaré inequality with constant 1. To show that this is optimal, simply consider the function $f(x) = x_1 + \cdots + x_n$.

Exercise 2.2 Prove the formula (1.4) of Chap. 1 using the covariance lemma and the explicit representation of the Ornstein-Uhlenbeck Dirichlet form.

5 Some Applications of the Gaussian Poincaré Inequality

Recall the $(1+1)$-dimensional Gaussian polymer model of length n from Chap. 1, Sect. 1.2. Let g_v be the Gaussian weight attached to vertex v and E_n be the ground state energy. It is not difficult to see that E_n is an absolutely continuous function of the vertex weights. It is also an easy fact, as mentioned above, that the Gaussian

Poincaré inequality (2.5) applies to absolutely continuous functions when the right-hand side is well-defined.

Now, varying a vertex weight g_v by a very small amount can affect E_n only if v belongs to the optimal path. More precisely (by an obvious abuse of notation),

$$\frac{\partial E_n}{\partial g_v} = -1_{\{v \in \text{optimal path}\}}.$$

Therefore,

$$|\nabla E_n|^2 = \sum_v \left(\frac{\partial E_n}{\partial g_v}\right)^2$$

$$= \sum_v 1_{\{v \in \text{optimal path}\}}$$

$$= \text{number of vertices in optimal path} = n+1.$$

Consequently, the Gaussian Poincaré inequality gives

$$\text{Var}(E_n) \leq n+1.$$

This is quite a remarkable result, considering that the expected size of E_n grows like a positive multiple of n (by subadditivity and a simple argument for the lower bound). Indeed, I do not know of a different way to prove this.

We end this section with two simple exercises for the reader. First, recall the free energy $F_n(\beta)$ of the SK model at inverse temperature β (Chap. 1, Sect. 1.3).

Exercise 2.3 Using the Poincaré inequality, prove that $\text{Var}(F_n(\beta)) \leq Cn$ where C is a universal constant that does not depend on β or n.

Next, recall the variance inequality (1.1) from Sect. 1.4 of Chap. 1.

Exercise 2.4 Suppose that $g_1, \ldots, g_n$ are jointly Gaussian random variables with arbitrary correlations. Prove that

$$\text{Var}\left(\max_{1 \leq i \leq n} g_i\right) \leq \max_{1 \leq i \leq n} \text{Var}(g_i).$$

6 Fourier Expansion

Let $(X_t)_{t \geq 0}$ be a reversible Markov process on some state space, with semigroup P_t, equilibrium measure μ, generator L and Dirichlet form $\mathcal{E}$. Since the process is reversible, L is self-adjoint. It is also true that L is negative semidefinite. To see this, note that by the Cauchy-Schwarz inequality and Jensen's inequality, for any $f \in L^2(\mu)$ we have

$$\int f P_t f \, d\mu \leq \left(\int f^2 \, d\mu \right)^{1/2} \left(\int (P_t f)^2 \, d\mu \right)^{1/2} \leq \int f^2 \, d\mu.$$

Therefore, by Fatou's lemma,

$$(f, Lf) = \int \lim_{t \to 0} \frac{f(P_t f - f)}{t} d\mu \leq \liminf_{t \to 0} \int \frac{f(P_t f - f)}{t} d\mu \leq 0.$$

Note that 0 is always an eigenvalue of L since L applied to any constant function gives zero. Often (and in particular for the OU process, as we will see later) the eigenvalues of $-L$ may be ordered as a countable sequence like $0 = \lambda_0 \leq \lambda_1 \leq \lambda_2 \leq \cdots$, with a corresponding sequence of orthonormal eigenfunctions $u_0, u_1, u_2, \ldots$ where $u_0 \equiv 1$. This sequence of eigenfunctions forms a complete orthogonal basis of $L^2(\mu)$, since if $Lf \equiv 0$, then $P_t f = e^{tL} f = f$ for all f, and therefore $f = \lim_{t \to \infty} P_t f \equiv \int f d\mu$. In particular, by the Plancherel identity,

$$\|f\|_{L^2(\mu)}^2 = \sum_{k=0}^{\infty} (u_k, f)^2.$$

Now $(u_0, f) = \mathbb{E}_\mu(f)$. Therefore the Plancherel identity may be alternatively written as

$$\mathrm{Var}_\mu(f) = \sum_{k=1}^{\infty} (u_k, f)^2.$$

Again, since

$$f = \sum_{k=0}^{\infty} (u_k, f) u_k,$$

it follows that

$$\mathcal{E}(f, f) = -(f, Lf) = \sum_{k=0}^{\infty} \lambda_k (u_k, f)^2 = \sum_{k=1}^{\infty} \lambda_k (u_k, f)^2. \tag{2.6}$$

In fact, the right-hand side may be taken as an alternative definition of the Dirichlet form when the spectrum is countable.

The following well-known lemma characterizes Poincaré inequalities in terms of the spectrum.

Lemma 2.5 *The Markov semigroup P_t satisfies a Poincaré inequality if and only if $\lambda_1 > 0$, and in that case the optimal constant is $1/\lambda_1$.*

Proof Suppose that $\lambda_1 > 0$. Then for any $f \in L^2(\mu)$,

$$\mathrm{Var}_\mu(f) = \sum_{k=1}^\infty (u_k, f)^2 \le \frac{1}{\lambda_1} \sum_{k=1}^\infty \lambda_k (u_k, f)^2 = \frac{1}{\lambda_1} \mathcal{E}(f, f).$$

To show that $1/\lambda_1$ is the optimal constant, simply observe that equality is achieved in the above display when $f = u_1$.

If $\lambda_1 = 0$, then u_1, being orthogonal to constant functions, must be non-constant and hence must have nonzero variance. On the other hand $\mathcal{E}(u_1, u_1) = 0$. This shows that the semigroup cannot satisfy a Poincaré inequality. □

Chapter 3
Superconcentration and Chaos

This chapter contains the formal definitions of superconcentration and chaos in a generalized setting, and the proof that they are equivalent.

1 Definition of Superconcentration

Let X_t be a reversible Markov process, with semigroup P_t, generator L, equilibrium measure μ and Dirichlet form $\mathcal{E}$. Suppose that the process satisfies a Poincaré inequality with optimal constant C.

Definition 3.1 In the above setting, we will say that a real valued function f on the state space of the Markov process X_t is ϵ-superconcentrated if

$$\mathrm{Var}_\mu(f) \le \epsilon\, C\, \mathcal{E}(f, f).$$

To understand this definition, we must think of ϵ being 'small', in the sense that there is some n lurking in the background, such that ϵ goes to zero as $n \to \infty$. When ϵ is small in this sense, we will simply say that f is superconcentrated instead of ϵ-superconcentrated.

As an example, consider the n-dimensional OU semigroup, whose equilibrium measure is the n-dimensional standard Gaussian measure γ^n. Let $f_1 : \mathbb{R}^n \to \mathbb{R}$ be the function

$$f_1(x) = \max_{1 \le i \le n} x_i.$$

Then it is well-known from the theory of extrema of Gaussian random variables (see Appendix A) that as $n \to \infty$,

$$\mathrm{Var}_{\gamma^n}(f_1) \sim \frac{C}{\log n}$$

for some positive constant C. But f_1 is absolutely continuous with $|\nabla f_1|^2 \equiv 1$, and so $\mathcal{E}(f_1, f_1) = 1$. Since the spectral gap of this semigroup is 1 irrespective of n (proved earlier in Chap. 2, Sect. 4), this shows that f_1 is ϵ_n-superconcentrated, where $\epsilon_n = C/\log n$.

On the other hand, the function $f_2(x) = x_1 + \cdots + x_n$ is not superconcentrated as $n \to \infty$ (and by that we mean that it is not ϵ_n-superconcentrated for any sequence $\epsilon_n \to 0$).

Suppose now that $g_1, \ldots, g_n$ are jointly Gaussian, but not necessarily independent. When is $\max_{1 \le i \le n} g_i$ superconcentrated? The general answer to this question is not known. One may feel that the maximum may be superconcentrated if the 'correlations are small'. While this intuition is probably correct, the question is 'how small'? To see the difficulty, consider the following. Let $g_1, \ldots, g_n$ be i.i.d. standard Gaussian random variables, and for each $\sigma \in \{-1, 1\}^n$ define

$$H(\sigma) := \frac{1}{\sqrt{n}} \sum_{i=1}^{n} g_i \sigma_i.$$

Then the collection $(H(\sigma))_{\sigma \in \{-1,1\}^n}$ is jointly Gaussian, with mean zero, variance 1, and

$$\mathrm{Cov}\big(H(\sigma), H(\sigma')\big) = \frac{1}{n} \sum_{i=1}^{n} \sigma_i \sigma'_i,$$

which is close to zero for 'most pairs' (σ, σ'). However,

$$\max_{\sigma \in \{-1,1\}^n} H(\sigma) = \frac{1}{\sqrt{n}} \sum_{i=1}^{n} |g_i|,$$

which is not superconcentrated. Thus, simply 'most correlations close to zero' does not guarantee superconcentration of the maximum.

On the other hand if we had a quadratic form like $\sum g_{ij} \sigma_i \sigma_j$ instead of a linear form, then the maximum would most likely be superconcentrated, but I don't know how to prove this.

Open Problem 3.2 Suppose that $(g_{ij})_{1 \le i < j \le n}$ is a collection of i.i.d. standard Gaussian random variables. Prove that

$$\max_{\sigma \in \{-1,1\}^n} \sum_{1 \le i < j \le n} g_{ij} \sigma_i \sigma_j$$

is ϵ_n-superconcentrated for some $\epsilon_n \to 0$ as $n \to \infty$.

The superconcentration of the maxima of Gaussian fields was the main focus of Chatterjee (2008b). We will discuss more about this later.

2 Superconcentration and Noise-Sensitivity

Suppose that $-L$ has a spectral decomposition with eigenvalues $0 = \lambda_0 < \lambda_1 \leq \lambda_2 \leq \cdots$ and eigenfunctions $u_0, u_1, \ldots$. Recall from Chap. 2, Sect. 6 that $\text{Var}_\mu(f) = \sum_{k=1}^\infty (u_k, f)^2$ and $\mathcal{E}(f, f) = \sum_{k=1}^\infty \lambda_k (u_k, f)^2$. This shows that f is superconcentrated if and only if "most of the Fourier mass concentrates on the higher end of the spectrum", to borrow a notion from the noise-sensitivity literature (see e.g. Kahn et al. 1988, Friedgut and Kalai 1996, Talagrand 1997, Friedgut 1998, 1999, Benjamini et al. 2001, Mossel et al. 2003, 2010, Garban et al. 2010 and references therein). What this means is that if we define a probability measure on the positive integers that puts mass proportional to $(u_k, f)^2$ at k, and if f is superconcentrated, then the expected value given by the probability measure is large; in other words, most of the mass concentrates on large values of k. This condition is exactly the same as the one that defines noise-sensitive Boolean functions. In this sense, the theory presented in this monograph has a direct contact with the theory of noise-sensitivity in computer science.

However, there is a basic difference between noise-sensitivity and our pursuit. In noise-sensitivity, one usually considers only Boolean functions, that is, functions that take values in $\{0, 1\}$. We are concerned with functions that can take any real value, and in particular the variance of such functions. Our notion of chaos (introduced in Chap. 1 and formally defined below) has similarities with noise-sensitivity of Boolean functions but is substantially different in that it deals with the sensitivity of certain *structures* (such as ground state polymers) to noise rather than rapid decay of correlations. The notions of superconcentration and multiple valleys do not occur in the noise-sensitivity literature.

3 Definition of Chaos

Let all notation be as in Sects. 1 and 2. Assume that $\lambda_1 > 0$. Then note that for any f and any $t \geq 0$,

$$\mathcal{E}(f, P_t f) = -\bigl(f, L e^{tL} f\bigr) = \sum_{k=1}^\infty \lambda_k e^{-\lambda_k t} (u_k, f)^2$$

$$\leq e^{-\lambda_1 t} \sum_{k=1}^\infty \lambda_k (u_k, f)^2 = e^{-\lambda_1 t} \mathcal{E}(f, f).$$

We will say that f is chaotic if the above inequality is suboptimal for all t above a small threshold. The definition is formalized below.

Definition 3.3 We will say that a function f is (ϵ, δ)-chaotic if for all $t \geq \delta$,

$$\mathcal{E}(f, P_t f) \leq \epsilon \, e^{-\lambda_1 t} \, \mathcal{E}(f, f).$$

As usual, think of ϵ and δ being 'small', in the sense that they depend on some n lurking in the background and tend to zero as $n \to \infty$. If this is the case, then we will simply say that f is chaotic.

Admittedly, this is a strange and opaque definition. The following example may clarify matters a bit.

Consider the $(1+1)$-dimensional Gaussian polymer model introduced in Chap. 1, Sect. 1.2. Let E_n be the ground state energy of the polymer. Let g_v be the weight of the vertex v and $(g'_v)_{v \in \mathbb{Z}^2}$ be i.i.d. standard Gaussian random variables, independent of g_v. Define

$$g_v^t := e^{-t} g_v + \sqrt{1 - e^{-2t}} g'_v.$$

Let $V := \{-n, -n+1, \ldots, n-1, n\}^2$ and $N := |V|$. Define $g := (g_v)_{v \in V}$ and $g^t := (g_v^t)_{v \in V}$. Let P_t be the N-dimensional OU semigroup and γ^N be the N-dimensional standard Gaussian measure. Then for any $f : \mathbb{R}^V \to \mathbb{R}$,

$$\mathbb{E}_{\gamma^N}(f P_t f) = \mathbb{E}\big(f(g) \mathbb{E}(f(g^t) \mid g)\big) = \mathbb{E}\big(f(g) f(g^t)\big).$$

As in Chap. 2, Sect. 5, observe that

$$\frac{\partial E_n}{\partial g_v} = -1_{\{v \in \text{ optimal path}\}}.$$

Let $\hat{p}^t$ be the optimal path in the environment g^t, and let $\hat{p}$ be the optimal path in the environment g (so that $\hat{p} = \hat{p}^0$).

From Chap. 2, Sect. 4, recall the expression for the Dirichlet form $\mathcal{E}$ for the OU process, and the fact that $\nabla P_t = e^{-t} P_t \nabla$. Then

$$\begin{aligned}
\mathcal{E}(E_n, P_t E_n) &= \mathbb{E}_{\gamma^N}(\nabla E_n \cdot \nabla P_t E_n) \\
&= e^{-t} \mathbb{E}_{\gamma^N}(\nabla E_n \cdot P_t \nabla E_n) \\
&= e^{-t} \mathbb{E}\big(\nabla E_n(g) \cdot \nabla E_n(g^t)\big) \\
&= e^{-t} \sum_{v \in V} \mathbb{P}\big(v \in \hat{p} \text{ and } v \in \hat{p}^t\big) \\
&= e^{-t} \mathbb{E}\big|\hat{p} \cap \hat{p}^t\big|.
\end{aligned}$$

For the OU process, the results of Sects. 5 and 6 in Chap. 2 combine to give that $\lambda_1 = 1$. Thus, to say that E_n is (ϵ, δ)-chaotic is the same as saying that for all $t \geq \delta$,

$$\mathbb{E}\big|\hat{p} \cap \hat{p}^t\big| \leq \epsilon (n+1).$$

If ϵ and δ tend to zero as $n \to \infty$, this means that for large n, a small perturbation in the environment results in a new optimal path (ground state) that is almost disjoint from the original one.

Exercise 3.4 In the SK model, let $R_{1,2}(t)$ be the overlap between two configurations, one drawn from the original Gibbs measure at inverse temperature β, and another from the t-perturbed Gibbs measure (see definition in Chap. 1, Sect. 2.2). Show that if the free energy $F_n(\beta)$ is (ϵ, δ)-chaotic, then for all $t \geq \delta$, $\mathbb{E}(R_{1,2}^2(t)) \leq \epsilon$. In other words, a small perturbation leads to the two configurations being almost orthogonal.

4 Equivalence of Superconcentration and Chaos

Let all notation be as in the first three sections of this chapter. Assume that $\lambda_1 > 0$. The following theorem shows that superconcentration and chaos are the same thing. This is one of the main results of this monograph. A preliminary version was proved in Chatterjee (2008b).

Theorem 3.5 (Generalization of Theorem 1.8 in Chatterjee 2008b) *Suppose that f is ϵ-superconcentrated. Then for any $\delta > 0$, f is (ϵ', δ)-chaotic, where*

$$\epsilon' = \frac{\epsilon \, e^{\lambda_1 \delta}}{\lambda_1 \delta}.$$

Conversely, suppose that f is (ϵ, δ)-chaotic. Then f is ϵ'-superconcentrated, where

$$\epsilon' = \epsilon + \lambda_1 \delta.$$

Note that if f is superconcentrated with small ϵ, and λ_1 is not large, then choosing $\delta = \sqrt{\epsilon}$ gives ϵ' and δ that are both small. Similarly, if f is (ϵ, δ)-chaotic for some small ϵ and δ, then $\epsilon' = \epsilon + \lambda_1 \delta$ is small too.

The basic idea behind Theorem 3.5 is the following: By the covariance lemma,

$$\mathrm{Var}_\mu(f) = \int_0^\infty \mathcal{E}(f, P_t f) \, dt.$$

Therefore, if chaos holds, i.e. if $\mathcal{E}(f, P_t f)$ goes down rapidly to zero as t increases, then $\mathrm{Var}_\mu(f)$ must be small, which is the basis of 'chaos implies superconcentration'. On the other hand by spectral representation of the Dirichlet form, one can show that $\mathcal{E}(f, P_t f)$ is a non-negative and decreasing function of t. This shows that for the integral to be small, $\mathcal{E}(f, P_t f)$ must go down rapidly to zero as t increases. This can be formalized to show that 'superconcentration implies chaos'.

Proof of Theorem 3.5 Suppose that f is ϵ-superconcentrated. Then by the covariance lemma (Lemma 2.1) and the characterization of the optimal constant in the Poincaré inequality in terms of the spectral gap (Chap. 2, Sect. 6),

$$\frac{\epsilon}{\lambda_1} \mathcal{E}(f, f) \geq \mathrm{Var}_\mu(f) = \int_0^\infty \mathcal{E}(f, P_t f) \, dt.$$

By (2.6) from Chap. 2, Sect. 6, and the fact that $P_t = e^{tL}$, it follows that

$$\mathcal{E}(f, P_t f) = \sum_{k=1}^{\infty} \lambda_k e^{-\lambda_k t} (u_k, f)^2. \tag{3.1}$$

Therefore in particular $\mathcal{E}(f, P_t f)$ is non-negative for all t and is a decreasing function of t. Thus for any $\delta > 0$,

$$\frac{\epsilon}{\lambda_1} \mathcal{E}(f, f) \geq \int_0^\delta \mathcal{E}(f, P_t) \, dt \geq \delta \, \mathcal{E}(f, P_\delta f).$$

Therefore

$$e^{\lambda_1 \delta} \mathcal{E}(f, P_\delta f) \leq \frac{\epsilon \, e^{\lambda_1 \delta}}{\lambda_1 \delta} \mathcal{E}(f, f).$$

But the formula (3.1) also shows that $e^{\lambda_1 t} \mathcal{E}(f, P_t f)$ is a non-increasing function of t. Combining this with the above inequality, we have that for all $t \geq \delta$,

$$e^{\lambda_1 t} \mathcal{E}(f, P_t f) \leq \frac{\epsilon \, e^{\lambda_1 \delta}}{\lambda_1 \delta} \mathcal{E}(f, f).$$

This proves one part of the theorem. For the other part, simply observe that if f is (ϵ, δ)-chaotic, then by the decreasing nature of $\mathcal{E}(f, P_t f)$,

$$\mathrm{Var}_\mu(f) = \int_0^\infty \mathcal{E}(f, P_t f) \, dt$$

$$\leq \int_0^\delta \mathcal{E}(f, f) \, dt + \int_\delta^\infty \epsilon \, e^{-\lambda_1 t} \mathcal{E}(f, f) \, dt$$

$$\leq \delta \mathcal{E}(f, f) + \frac{\epsilon}{\lambda_1} \mathcal{E}(f, f),$$

which proves that f is $(\epsilon + \lambda_1 \delta)$-superconcentrated. □

5 Some Applications of the Equivalence Theorem

As an application of the equivalence theorem (Theorem 3.5), we will now see how to prove the equivalence of chaos and superconcentration of the ground state energy in the polymer model.

Proof of Theorem 1.10 of Chap. 1 Recall the Gaussian polymer model defined in Chap. 1, Sect. 1.2. Let E_n denote the ground state energy. Let $\hat{p}_n$ denote the optimal path of length n and $\hat{p}_n^t$ denote the optimal path of length n in the t-perturbed environment. From the discussion in Chap. 2, Sect. 5, it is clear that $\mathrm{Var}(E_n) = o(n)$ if and only if E_n is ϵ_n-superconcentrated for some $\epsilon_n \to 0$ as $n \to \infty$. On the other hand, from the discussion in Sect. 3 of the present chapter it follows that E_n is (ϵ_n, δ_n)-chaotic for some ϵ_n and δ_n tending to zero if and only if there exists a sequence $t_n \to 0$ such that $\mathbb{E}|\hat{p}_n \cap \hat{p}_n^{t_n}| = o(n)$. By Theorem 3.5, this completes the proof of Theorem 1.10. □

The equivalence of chaos and superconcentration in the SK model and the maxima of Gaussian fields are left as exercises.

Exercise 3.6 Using the equivalence theorem, prove the equivalence of chaos and superconcentration of the free energy in the SK model (Theorem 1.12 of Chap. 1).

Exercise 3.7 Using the technique of proof of the equivalence theorem, prove that the maximum of a Gaussian field on a finite index set is superconcentrated if and only if the location of the maximum is chaotic under small perturbations (Theorem 1.13 of Chap. 1).

6 Chaotic Nature of the First Eigenvector

A random Hermitian matrix $A = (a_{ij})_{1 \leq i,j \leq n}$ is said to belong to the Gaussian Unitary Ensemble (GUE) if (i) $(a_{ij})_{1 \leq i \leq j \leq n}$ are independent random variables, (ii) the diagonal entries are standard real Gaussian random variables, and (iii) $(a_{ij})_{1 \leq i < j \leq n}$ are standard complex Gaussian random variables (i.e. real and imaginary parts are i.i.d. $N(0, 1/2)$).

Eigenvalues of GUE matrices are among the most widely studied objects in random matrix theory. For a general introduction to the classical random matrix ensembles and results, we refer to the book by Mehta (1991). The study of the largest eigenvalue was revolutionized through the work of Tracy and Widom (1994a,b, 1996). One of the striking implications of their work is that the largest eigenvalue has variance of order $n^{-1/3}$, beating the $O(1)$ bound given by standard isoperimetric and martingale methods. But the Tracy-Widom result is in the sense of weak convergence, and does not provide an actual bound on the variance that we need. A variance bound of order $n^{-1/3}$ follows from the works of Ledoux (2003, 2007) and Aubrun (2005).

The eigenvectors of GUE matrices, taken as rows (or columns) of a matrix, give rise to a Haar-distributed unitary matrix. In that sense, they are quite well-understood. However, the behavior of the eigenvectors under perturbations of the matrix has not been studied. Such questions arise, for instance, in the study of chaos in the spherical Sherrington-Kirkpatrick model of spin glasses. For the definition of this model and further references, let us refer to the paper of Panchenko and Talagrand (2007), where it was proved that the model is chaotic with respect to an external field. The goal of this section is to show that the first eigenvector is unstable under small perturbations of the matrix, and give a quantitative version of this statement. In the spherical SK model with complex spins, this establishes chaos with respect to the disorder at zero temperature.

Let us now formulate the question in terms of Gaussian fields. For each vector $u = (u_1, \ldots, u_n)$ in the unit sphere S^{2n-1} of $\mathbb{C}^n$, define the quadratic form

$$X_u := u^* A u,$$

where, as usual, u^* is the adjoint (i.e. conjugate transposed) of u. Since A is Hermitian, X_u is a real Gaussian random variable.

Now, if $v = zu$ for some z in the unit sphere $U(1)$ of $\mathbb{C}$, then $X_v \equiv X_u$. Therefore to retain identifiability, we define X_u for each u not in S^{2n-1}, but in the complex projective space $\mathbb{CP}^{n-1} = S^{2n-1}/U(1)$. However, we will continue to write elements of $\mathbb{CP}^{n-1}$ as if they were elements of $\mathbb{C}^n$, with the quotienting being implicit. With that convention, let

$$\lambda_1 := \max_{u \in \mathbb{CP}^{n-1}} X_u, \qquad u_1 := \operatorname{argmax}_{u \in \mathbb{CP}^{n-1}} X_u.$$

Then λ_1 is the largest eigenvalue of the GUE matrix A and u_1 is the corresponding unit eigenvector. Our objective in this section is to show that u_1 is chaotic under small perturbations of A.

Here we must remark that u_1 is almost surely well-defined in $\mathbb{CP}^{n-1}$. This follows from the fact that the eigenvalues all have multiplicity 1 almost surely, which can be deduced, for instance, from the well-known joint density of the eigenvalues of GUE (see e.g. Mehta 1991, Chap. 3).

Now let A' be an independent copy of A, and as usual define the perturbed matrix $A^t := e^{-t} A + \sqrt{1 - e^{-2t}} A'$. Let u_1^t be the first eigenvector of A^t. We want to show that $|u_1 \cdot u_1^t|$ tends to decay rapidly with t. Here $x \cdot y$ denotes the usual Euclidean inner product between vectors in $\mathbb{C}^n$. Note that there is no ambiguity because for any $u, v \in \mathbb{CP}^{n-1}$, $|u \cdot v|$ is well-defined (although $u \cdot v$ is not).

Theorem 3.8 *There is a universal constant C such that for any $t \geq 0$,*

$$\mathbb{E}|u_1 \cdot u_1^t|^2 \leq \frac{C}{(1 - e^{-t}) n^{1/3}},$$

where u_1^t is the first eigenvector of the perturbed matrix A^t defined above.

Remarks This shows that whenever $t \gg n^{-1/3}$, the vectors u_1 and u_1^t are almost orthogonal with high probability. As mentioned before, this result also proves that the ground state of the (complex) spherical Sherrington-Kirkpatrick model is chaotic under small perturbations of the disorder.

Proof An easy computation gives that for any $u, v \in \mathbb{CP}^{n-1}$,

$$\operatorname{Cov}(X_u, X_v) = |u \cdot v|^2,$$

where one should note that the right-hand side is well-defined on $\mathbb{CP}^{n-1}$. It is known from random matrix theory (see Ledoux (2003, 2007) and Aubrun (2005)) that

Var(λ_1) $\leq Cn^{-1/3}$. (We will see a proof of this in Sect. 1 of Chap. 9.) Therefore the above formula for the covariance and Theorem 1.13 seem to imply that the proof is done. However, we have to be a little careful because Theorem 1.13 works only for Gaussian fields on finite index sets. But this can be easily taken care of by taking the Gaussian field (X_u) restricted to finer and finer nets of points in $\mathbb{CP}^{n-1}$ and using the uniqueness of the maximizer and continuity to pass to the limit. □

Chapter 4
Multiple Valleys

The notion of multiple valleys has already been formally defined in Chap. 1, Sect. 3.2. In this chapter, we will give a general sketch as to why chaos (or equivalently, superconcentration) implies the existence of multiple valleys. The general idea will then be applied to the special examples of Gaussian polymers and the SK model.

1 Chaos Implies Multiple Valleys: The General Idea

I will now give an outline as to how chaos implies multiple valleys in random optimization problems, by considering the specific case of $(d+1)$-dimensional Gaussian random polymers. This argument appeared for the first time in Chatterjee (2008b), and was extended in Chatterjee (2009).

Consider the $(d+1)$-dimensional Gaussian random polymer model of length n. Let $g = (g_v)_{v \in \mathbb{Z}^d}$ be the environment, E_n be the ground state energy, and H_n be the energy function (Hamiltonian). Our goal is to show that if n is large and the ground state energy is superconcentrated (equivalently, chaotic), then with high probability, there exists many paths with near-minimal energy that are all nearly disjoint from each other.

The general sketch of the argument is as follows.

(1) Let H_n be the energy function, and let H'_n denote a small perturbation of the energy function (by replacing vertex weights with a new set of weights that have correlation ≈ 1 with the old weights, as per our usual practice).
(2) Let $\hat{p}$ denote the optimal (i.e. minimum energy) path for the energy function H_n and $\hat{p}'$ denote the optimal path for energy H'_n.
(3) By the chaos property, $\hat{p}$ and $\hat{p}'$ are nearly disjoint.
(4) Since H'_n is a small perturbation of H_n, it follows that $H'_n(\hat{p}') \approx H_n(\hat{p}')$.
(5) By the concentration of the ground state energy, $H'_n(\hat{p}') \approx$ the expected value of the ground state energy $\approx H_n(\hat{p})$.

S. Chatterjee, *Superconcentration and Related Topics*,
Springer Monographs in Mathematics, DOI 10.1007/978-3-319-03886-5_4,
© Springer International Publishing Switzerland 2014

(6) Combining the last two observations gives $H_n(\hat{p}') \approx H_n(\hat{p})$. In other words, $\hat{p}'$ is a near-optimal path for the energy function H_n.
(7) Therefore, we have found two paths $\hat{p}$ and $\hat{p}'$ that are both near-optimal for the energy function H_n, and are nearly disjoint. Repeating this process gives many such paths.

In the next section this argument is made rigorous.

2 Multiple Valleys in Gaussian Polymers

The following theorem fills in the details of the argument presented in the preceding section and thereby proves that if the ground state energy of the $(d+1)$-dimensional Gaussian polymer (defined in Chap. 1, Sect. 1.2) is chaotic, then there are multiple valleys in the energy landscape. We will later prove that the ground state energy in the $(1+1)$-dimensional model is superconcentrated. Together with the equivalence of chaos and superconcentration, this will prove the multiple valley theorem for Gaussian polymers (Theorem 1.17).

Theorem 4.1 *Consider the $(d+1)$-dimensional Gaussian random polymer, $d \geq 2$. Suppose that the ground state energy E_n is (ϵ_n, δ_n)-chaotic, where both ϵ_n and δ_n tend to zero as $n \to \infty$. Then the energy function of the model has multiple valleys as $n \to \infty$.*

Proof Let K_n be a sequence of numbers tending to infinity, to be chosen later. Fix n, and let $h^1, \ldots, h^{K_n}$ be independent copies of the environment $g = (g_v)_{v \in \mathbb{Z}^d}$. For each $1 \leq i \leq K_n$ and $v \in \mathbb{Z}^d$, let

$$g_v^i := e^{-\delta_n/2} g_v + \sqrt{1 - e^{-\delta_n}} h_v^i.$$

Then notice that for each $i \neq j$ and each v, the correlation between g_v^i and g_v^j is $e^{-\delta_n}$. In particular, the environment g^j is a δ_n-perturbation of the environment g^i. Therefore by the chaos property, if $\hat{p}^i$ denotes the optimal path of length n in environment g^i, then

$$\mathbb{E}|\hat{p}^i \cap \hat{p}^j| \leq \epsilon_n(n+1).$$

Consequently,

$$\mathbb{E}\left(\max_{1 \leq i \neq j \leq K_n} |\hat{p}^i \cap \hat{p}^j|\right) \leq \mathbb{E}\left(\sum_{1 \leq i \neq j \leq K_n} |\hat{p}^i \cap \hat{p}^j|\right) \leq C K_n^2 \epsilon_n n, \quad (4.1)$$

where C denotes, here and below, any constant that does not depend on n.

Let H^i denote the energy function for the environment g^i and let H be the energy function for the environment g. (We drop the subscript n from H_n for notational simplicity.) Then note that for any path p of length n, and any i,

$$|H^i(p) - H(p)| \le \left(1 - e^{-\delta_n/2}\right)\left|\sum_{v \in p} g_v\right| + \sqrt{1 - e^{-\delta_n}}\left|\sum_{v \in p} h^i_v\right|$$

$$\le \left(1 - e^{-\delta_n/2}\right) \max_q \left|\sum_{v \in q} g_v\right| + \sqrt{1 - e^{-\delta_n}} \max_q \left|\sum_{v \in q} h^i_v\right|,$$

where q runs over all paths of length n. Since the above maxima have expected value bounded by Cn by simple properties of Gaussian random variables (as in the results contained in Appendix A), this shows that

$$\mathbb{E}\left(\max_p |H^i(p) - H(p)|\right) \le C\sqrt{\delta_n} n.$$

In particular,

$$\mathbb{E}|H^i(\hat{p}^i) - H(\hat{p}^i)| \le C\sqrt{\delta_n}\, n. \qquad (4.2)$$

Let $\hat{p}$ denote the optimal path in the environment g. Then note that for all i, $H^i(\hat{p}^i)$ has the same distribution as $H(\hat{p})$, and that $H(\hat{p}^i) \ge H(\hat{p})$. Therefore by (4.2),

$$\mathbb{E}|H(\hat{p}^i) - H(\hat{p})| = \mathbb{E}(H(\hat{p}^i) - H(\hat{p}))$$
$$= \mathbb{E}(H(\hat{p}^i) - H^i(\hat{p}^i)) \le C\sqrt{\delta_n}\, n.$$

Consequently,

$$\mathbb{E}\left(\max_{1 \le i \le K_n} |H(\hat{p}^i) - H(\hat{p})|\right) \le CK_n\sqrt{\delta_n}\, n. \qquad (4.3)$$

Now choose K_n to be growing to infinity so slowly that as $n \to \infty$,

(1) $K_n^2 \epsilon_n \to 0$, and
(2) $K_n\sqrt{\delta_n} \to 0$.

Then from (4.1) and (4.3) we conclude that there is a sequence η_n going to zero slowly enough such that

$$\lim_{n \to \infty} \mathbb{P}\left(\max_{1 \le i \ne j \le K_n} |\hat{p}^i \cap \hat{p}^j| > \eta_n n\right) = 0$$

and

$$\lim_{n \to \infty} \mathbb{P}\left(\max_{1 \le i \le K_n} |H(\hat{p}^i) - H(\hat{p})| > \eta_n n\right) = 0.$$

Combining this with the fact that $\liminf_{n \to \infty} |H(\hat{p})|/n > 0$ (left as an exercise to the reader, to be proved using results from Appendix A), this completes the proof that chaos implies multiple valleys in the $(d+1)$-dimensional Gaussian polymer model. $\square$

3 Multiple Valleys in the SK Model

Note that the general line of argument given in Sect. 1 of this chapter has nothing to do with polymers in particular. The same applies to various other models, including the Sherrington-Kirkpatrick model. However one problem with the SK model is that the ground state energy has not yet been proven to be superconcentrated. We will circumvent this problem by showing that if the free energy is superconcentrated at all finite values of β (the inverse temperature), then the energy landscape has multiple valleys. This is accomplished by following more or less the same line of argument as in the proof of Theorem 4.1, and slowing taking β to infinity with n. This method works since the Gibbs measure is concentrated near the ground state when β is large.

Theorem 4.2 *Suppose that for all $\beta > 0$, there exists $\epsilon_n = \epsilon_n(\beta)$ and $\delta_n = \delta_n(\beta)$ tending to zero as $n \to \infty$, such that the free energy $F_n(\beta)$ of the SK model at inverse temperature β is (ϵ_n, δ_n)-chaotic. Then the energy landscape of the SK model has multiple valleys.*

Proof Throughout this proof, C will denote any constant that does not depend on any of the parameters.

Fix β_n and n. We will later choose β_n depending on n. As usual, let $R_{1,2}(t)$ denote the overlap between a configuration drawn from the Gibbs measure at inverse temperature β_n and another one from the t-perturbed measure. (See Chap. 1, Sect. 2.2 for the relevant definitions.)

From the hypothesis of the theorem, we have that $F_n(\beta_n)$ is (ϵ_n, δ_n)-chaotic, where ϵ_n and δ_n are determined by β_n. By Exercise 3.4 in Chap. 3, Sect. 3, this means that for all $t \geq \delta_n$, $\mathbb{E}(R_{1,2}^2(t)) \leq \epsilon_n$. Since $\lim_{n \to \infty} \epsilon_n(\beta) = 0$ and $\lim_{n \to \infty} \delta_n(\beta) = 0$ for each fixed β, one can choose a sequence $\beta_n \to \infty$ so slowly that $\epsilon_n(\beta_n) \to 0$ and $\delta_n(\beta_n) \to 0$. From now on, assume that β_n has been chosen to satisfy this criterion.

Let $g = (g_{ij})_{1 \leq i < j \leq n}$ denote the disorder. Let K_n be a sequence of numbers tending to infinity, to be chosen later. Fix n, and let $h^1, \ldots, h^{K_n}$ be independent copies of the disorder g. For each $1 \leq k \leq K_n$ and $1 \leq i < j \leq n$, let

$$g_{ij}^k := e^{-\delta_n/2} g_{ij} + \sqrt{1 - e^{-\delta_n}} h_{ij}^k.$$

Then notice that for each $1 \leq k \neq l \leq K_n$ and each $1 \leq i < j \leq n$, the correlation between g_{ij}^k and g_{ij}^l is $e^{-\delta_n}$. In particular, the environment g^l is a δ_n-perturbation of the environment g^k. Therefore by the chaos property, if $R_{k,l}$ denotes the overlap between a configuration σ^k picked from the Gibbs measure defined by the disorder g^k and another one σ^l from the Gibbs measure defined by g^l, then $\mathbb{E}(R_{k,l}^2) \leq \epsilon_n$. Consequently,

$$\mathbb{E}\left(\max_{1 \leq k \neq l \leq K_n} R_{k,l}^2\right) \leq \mathbb{E}\left(\sum_{1 \leq k \neq l \leq K_n} R_{k,l}^2\right) \leq K_n^2 \epsilon_n. \qquad (4.4)$$

3 Multiple Valleys in the SK Model

Let H^k denote the energy function for the disorder g^k, and H denote the energy function for the original disorder g. (As in the proof of Theorem 4.1 we drop the subscript n from H_n.) Then note that for any configuration τ and any k,

$$|H^k(\tau) - H(\tau)|$$

$$\leq \left(1 - e^{-\delta_n/2}\right) \left|\frac{1}{\sqrt{n}} \sum_{1 \leq i < j \leq n} g_{ij} \tau_i \tau_j\right| + \sqrt{1 - e^{-\delta_n}} \left|\frac{1}{\sqrt{n}} \sum_{1 \leq i < j \leq n} h_{ij}^k \tau_i \tau_j\right|$$

$$\leq \left(1 - e^{-\delta_n/2}\right) \max_{\tau' \in \{-1,1\}^n} \left|\frac{1}{\sqrt{n}} \sum_{1 \leq i < j \leq n} g_{ij} \tau_i' \tau_j'\right|$$

$$+ \sqrt{1 - e^{-\delta_n}} \max_{\tau' \in \{-1,1\}^n} \left|\frac{1}{\sqrt{n}} \sum_{1 \leq i < j \leq n} h_{ij}^k \tau_i' \tau_j'\right|.$$

Since the above maxima have expected value bounded by Cn by simple properties of Gaussian random variables (see Appendix A), this shows that

$$\mathbb{E}\left(\max_{\tau \in \{-1,1\}^n} |H^k(\tau) - H(\tau)|\right) \leq C\sqrt{\delta_n} n.$$

In particular,

$$\mathbb{E}|H^k(\sigma^k) - H(\sigma^k)| \leq C\sqrt{\delta_n} n. \tag{4.5}$$

Define $G_n(\beta) := -\beta F_n(\beta)$. Let

$$M := \min_{\tau \in \{-1,1\}^n} H(\tau).$$

Then for any β,

$$\beta M = -\log e^{-\beta M} \geq -\log \sum_\sigma e^{-\beta H(\sigma)} = -G_n(\beta)$$

$$\geq -\log(2^n e^{-\beta M}) = -n \log 2 + \beta M.$$

Thus,

$$|G_n(\beta) + \beta M| \leq n \log 2. \tag{4.6}$$

Let σ denote a configuration drawn from the Gibbs measure defined by the disorder g at inverse temperature β. Recall the notation $\langle \cdot \rangle$ that denotes averaging with respect to Gibbs measures. An easy verification shows that

$$G_n''(\beta) = \beta(\langle H(\sigma)^2 \rangle - \langle H(\sigma) \rangle^2) \geq 0.$$

Therefore G_n' is an increasing function of β and hence by (4.6),

$$G_n'(\beta) \geq \frac{2}{\beta} \int_{\beta/2}^{\beta} G_n'(x) dx$$

$$= \frac{G_n(\beta) - G_n(\beta/2)}{\beta/2} \geq -M - \frac{4n \log 2}{\beta}.$$

Combining this with the observation that $G'_n(\beta) = -\langle H(\sigma) \rangle$, we have that for any β,

$$\mathbb{E}\big(H(\sigma) - M\big) = -\mathbb{E}\big(G'_n(\beta) + M\big) \leq \frac{4n \log 2}{\beta}. \tag{4.7}$$

Now take $\beta = \beta_n$. Note that the random variable $H^k(\sigma^k)$ has the same (unconditional) distribution as $H(\sigma)$. Note also that $H(\sigma^k) \geq M$ for all k. These facts and the inequalities (4.5) and (4.7) imply that for any k,

$$\mathbb{E}\big|H(\sigma^k) - M\big| = \mathbb{E}\big(H(\sigma^k) - M\big)$$
$$= \mathbb{E}\big(H(\sigma^k) - H^k(\sigma^k)\big) + \mathbb{E}\big(H^k(\sigma^k) - M\big)$$
$$\leq C\sqrt{\delta_n}n + \frac{4n \log 2}{\beta_n}.$$

Thus,

$$\mathbb{E}\Big(\max_{1 \leq k \leq K_n} \big|H(\sigma^k) - M\big|\Big) \leq C\sqrt{\delta_n} K_n n + \frac{CK_n n}{\beta_n}. \tag{4.8}$$

We have already chosen β_n tending to ∞ so slowly that $\epsilon_n \to 0$ and $\delta_n \to 0$. Now choose $K_n \to \infty$ so slowly that

(1) $K_n^2 \epsilon_n \to 0$, and
(2) $\sqrt{\delta_n} K_n + K_n/\beta_n \to 0$.

Then from (4.4) and (4.8) we conclude that there is a sequence η_n going to zero slowly enough such that

$$\lim_{n \to \infty} \mathbb{P}\Big(\max_{1 \leq k \neq l \leq K_n} R_{k,l}^2 > \eta_n\Big) = 0$$

and

$$\lim_{n \to \infty} \mathbb{P}\Big(\max_{1 \leq k \leq K_n} \big|H(\sigma^k) - M\big| > \eta_n n\Big) = 0.$$

Combining this with the fact that $\liminf_{n \to \infty} |M|/n > 0$ (again, left as an exercise to the reader), this completes the proof that chaos implies multiple valleys in the SK model. □

4 Multiple Peaks in Gaussian Fields

In this section we prove Theorem 1.18 of Chap. 1, that is, superconcentration (or chaos) implies multiple peaks in Gaussian fields. The idea of the proof is basically a generalization of the idea of the proof of Theorem 4.1, with a few minor twists.

4 Multiple Peaks in Gaussian Fields

Proof of Theorem 1.18 of Chap. 1 Since we are allowed to increase the value of C without affecting the conclusion, it is okay to assume that ϵ is sufficiently small wherever required.

By the equivalence theorem (Theorem 3.5), it follows that the maximum is (ϵ', δ)-chaotic, where $\delta = \sqrt{\epsilon}$ and $\epsilon' = \sqrt{\epsilon} e^{\sqrt{\epsilon}}$. Since $\epsilon' \geq \delta$, this means that the maximum is (ϵ', ϵ')-chaotic.

Let g' be an independent copy of g and let $g^t := e^{-t}g + \sqrt{1-e^{-2t}}g'$ for each $t \geq 0$. Let $\gamma := (2\epsilon')^{1/4}$ and let k be an integer between $1/\gamma$ and $2/\gamma$. Let $h^1, \ldots, h^k$ be k i.i.d. copies of g. For $i = 1, \ldots, k$, define

$$w^i := e^{-\epsilon'}g + \sqrt{1-e^{-2\epsilon'}}h^i.$$

Let J^i be the index at which w^i is maximized. It is easy to see that for every $1 \leq i < j \leq k$, the pair (w^i, w^j) has the same distribution as that of $(g, g^{2\epsilon'})$. Since g is (ϵ', ϵ')-chaotic, this shows that for every $1 \leq i < j \leq k$,

$$\mathbb{E}(R(J^i, J^j)) \leq \epsilon.$$

Therefore, since all correlations are positive,

$$\mathbb{P}\left(\max_{1 \leq i < j \leq k} R(J^i, J^j) > \gamma\right) \leq \sum_{1 \leq i < j \leq k} \mathbb{P}(R(J^i, J^j) > \gamma) \leq \frac{k^2 \epsilon'}{2\gamma} \leq \gamma. \quad (4.9)$$

Now let

$$u^i := \sqrt{1-e^{-2\epsilon'}}g - e^{-\epsilon'}h^i.$$

Then u^i and w^i are independent, and

$$g = e^{-\epsilon'}w^i + \sqrt{1-e^{-2\epsilon'}}u^i. \quad (4.10)$$

Let $m := \mathbb{E}(\max_{1 \leq l \leq n} g_l)$ and

$$\eta := \frac{K}{\sqrt{\log(1/\gamma)}},$$

where K is a constant to be chosen later. Fix i and define three events E_1, E_2 and E_3 as:

$$E_1 := \left\{\max_{1 \leq l \leq n} w_l^i \geq (1-\eta)m\right\},$$

$$E_2 := \{u_{J^i}^i \geq -\eta m\},$$

$$E_3 := \left\{\max_{1 \leq l \leq n} g_l \leq (1+\eta)m\right\}.$$

If E_1, E_2 and E_3 happen and ϵ' is sufficiently small, then by (4.10),

$$g_{Ji} \geq e^{-\epsilon'}(1-\eta)m - \eta m$$
$$\geq (1-3\eta)m \geq (1-4\eta) \max_{1 \leq l \leq n} g_l.$$

By the inequality (A.7), $\mathbb{P}(E_1^c)$ and $\mathbb{P}(E_3^c)$ are both bounded by $e^{-\eta^2 m^2/2}$. By the independence of w and u, the inequality (A.2) from Appendix A, and the assumption that $R(i,i) = 1$ for all i, $\mathbb{P}(E_2^c)$ is bounded by $e^{-\eta^2 m^2/2}$. Combining all of the above observations, we see that

$$\mathbb{P}\left(g_{Ji} \geq (1-4\eta) \max_{1 \leq l \leq n} g_l \text{ for all } i\right) \geq 1 - 3ke^{-\eta^2 m^2/2}. \quad (4.11)$$

Now, by (4.9), there exists a subset S of $\{1, \ldots, n\}$ of size k such that $R(i,j) \leq \gamma$ for all distinct $i, j \in S$. By Sudakov minoration (Lemma A.3 in Appendix A) and the definitions of k and γ, this shows that

$$m \geq c \log(1/\gamma)$$

for some universal constant c. Using this lower bound in (4.11), combining with (4.9), and choosing K sufficiently large proves the theorem. □

5 Multiple Peaks in the NK Fitness Landscape

Kauffman and Levin (1987) introduced a class of models for the evolution of hereditary systems, which has since become one of the most popular models in evolutionary biology and some other areas. They named it the *NK* model because there are two parameters, N and K. The model envisions a genome as consisting of N genes, each of which exists as one of two possible alleles. The fitness score of an allele at a given site is determined by the alleles of the K neighboring sites. Other than that, the fitnesses are as simple as possible, namely i.i.d., and the fitnesses of different sites are averaged to get the overall fitness of a genome.

Let us define things formally. The space of all genomes is $\{0,1\}^N$. Let

$$Y(i;\eta), \quad 0 \leq i \leq N-1, \ \eta \in \{0,1\}^{K+1}$$

be a collection of i.i.d. random variables, assumed to be standard Gaussian for our purposes. Given $\sigma \in \{0,1\}^N$, define the 'fitness' of σ as

$$F(\sigma) := \sum_{i=0}^{N-1} Y\bigl(i; (\sigma_i, \sigma_{i+1}, \ldots, \sigma_{i+K})\bigr),$$

where the subscripts of σ are wrapped around, i.e. $\sigma_{N+i} = \sigma_i$. The function $F : \{0,1\}^N \to \mathbb{R}$ is called the 'fitness landscape', and the main objects of interest are the local and global maxima of this landscape.

5 Multiple Peaks in the *NK* Fitness Landscape

The *NK* model has been extensively studied, but very little of it is rigorous. The first rigorous paper on the model, to the best of our knowledge, was due to Evans and Steinsaltz (2002). These authors used elegant arguments involving max-plus algebras to carry out computations concerning the global and local maxima of the *NK* model when K is fixed and $N \to \infty$. One can also find in Evans and Steinsaltz (2002) a beautifully written introduction to the history and motivation behind the model. Further rigorous results were derived using different tools by Durrett and Limic (2003) who proved, among other things, a central limit theorem for the maximum fitness when K is fixed and $N \to \infty$. Some results on the local maxima in the case where N and K both tend to infinity were obtained by Limic and Pemantle (2004).

Our goal is to show that when N and K both tend to infinity, the fitness landscape has many global near-maxima, which are 'far away' from each other. In other words, there are many nearly globally fittest genomes that are drastically different from each other. To formalize this statement, we first need define what is meant by 'far away' in the space of genomes. The *NK* model naturally defines the following measure of proximity between two genomes σ and σ':

$$p_{N,K}(\sigma, \sigma') := |\{i : (\sigma_i, \ldots, \sigma_{i+K}) = (\sigma'_i, \ldots, \sigma'_{i+K})\}|. \quad (4.12)$$

This is a natural definition because

$$p_{N,K}(\sigma, \sigma') = \mathrm{Cov}(F(\sigma), F(\sigma')).$$

Note also that if K is large, then σ and σ' may be far apart even if the Hamming distance between σ and σ' is relatively small.

To understand the nature of the global maximum, we first have to have an idea about its size. It was shown by Evans and Steinsaltz (2002) that the size of the global maximum grows linearly in N when K is fixed (in an asymptotic sense). Following their argument, one can further deduce the surprising fact that when divided by N the expected size of the global maximum can be bounded above and below by universal constants that do not depend on K.

Lemma 4.3 *Irrespective of the value of K, we have*

$$\frac{N}{\sqrt{\pi}} \leq \mathbb{E}\left(\max_{\sigma} F(\sigma)\right) \leq N\sqrt{2\log 2}.$$

Remark In fact, the bounds are sharp. The lower bound is achieved when $K = 0$, and the upper bound is achieved (asymptotically) when $K = N - 1$. Moreover, as we will see in the proof, the expected value is an increasing function of K.

Proof The upper bound is straightforward from inequality (A.3) of Appendix A. For the lower bound, first observe that if $G(\sigma)$ is another measure of fitness, corre-

sponding to the NK model with $K = 0$, then for every σ, σ', we have
$$\mathrm{Cov}(G(\sigma), G(\sigma')) \ge \mathrm{Cov}(F(\sigma), F(\sigma')),$$
since it is quite clear that $p_{N,K}(\sigma, \sigma')$ is a decreasing function of K. Moreover $\mathrm{Var}(F(\sigma)) = \mathrm{Var}(G(\sigma)) = N$ for every σ. Therefore by Slepian's lemma,
$$\mathbb{E}\Big(\max_\sigma F(\sigma)\Big) \ge \mathbb{E}\Big(\max_\sigma G(\sigma)\Big).$$
Now, if $(Z(i, \eta))_{0 \le i \le N-1,\ \eta \in \{0,1\}}$ are the i.i.d. Gaussian fitness scores used to define G, then
$$\max_\sigma G(\sigma) = \sum_{i=0}^{N-1} \max\{Z(i,0), Z(i,1)\}.$$
It is easy to compute that the expected value of the maximum of two independent standard Gaussian random variables is $\pi^{-1/2}$. This completes the proof. □

The following theorem is the main result of this section. It shows that when N and K both tend to infinity, the fitness landscape exhibits the multiple peaks property. We first establish superconcentration using hypercontractivity, and then use Theorem 1.18 of Chap. 1 to deduce the existence of multiple peaks.

Theorem 4.4 *Suppose $N > K \ge 1$. Then $\mathrm{Var}(\max_\sigma F(\sigma)) \le CN/K$, where C is a universal constant. As a consequence, there is another universal constant C' such that if $\delta := C'(\log K)^{-1/2}$, then with probability at least $1 - \delta$ there exists $A \subseteq \{0, 1\}^N$ such that*

(a) $|A| \ge 1/\delta$,
(b) $p_{N,K}(\sigma, \sigma') \le \delta N$ *for all* $\sigma, \sigma' \in A$, $\sigma \ne \sigma'$, *and*
(c) *for all* $\sigma \in A$,
$$F(\sigma) \ge (1 - \delta) \max_{\sigma' \in \{0,1\}^N} F(\sigma').$$

Proof Let $M = \max_\sigma F(\sigma)$. Let $\hat{\sigma}$ be the maximizing configuration. With an obvious abuse of notation, we have
$$\frac{\partial M}{\partial Y(i;\eta)} = \mathbb{I}_{\{(\hat{\sigma}_i,\ldots,\hat{\sigma}_{i+K})=(\eta_1,\ldots,\eta_{K+1})\}}.$$
By the symmetry of the problem, this shows that
$$\left\|\frac{\partial M}{\partial Y(i;\eta)}\right\|_{L^1} = \left\|\frac{\partial M}{\partial Y(i;\eta)}\right\|_{L^2}^2 = 2^{-K-1}.$$
By Talagrand's L^1–L^2 theorem (Theorem 5.1), this immediately shows that $\mathrm{Var}(M) \le CN/K$. The rest follows from Theorem 1.18 of Chap. 1, taking $\epsilon = 1/K$. □

5 Multiple Peaks in the NK Fitness Landscape

Theorem 4.4 proves that when N and K are both large, there are many near-fittest genomes that are very different from each other. However, the following question of biological interest is open. This was suggested by Steve Evans.

Open Problem 4.5 In the NK model when N and K are both large, prove or disprove that the near-fittest genomes are connected by sequences of near-fittest genomes that form 'bridges', in the sense that any two neighboring members in a sequence are close to each other in the $p_{N,K}$ metric.

Chapter 5
Talagrand's Method for Proving Superconcentration

Until now, we have defined superconcentration, chaos and multiple valleys; and we have proved that superconcentration and chaos are equivalent, and that chaos implies multiple valleys. However, to get any mileage out of the interrelations between these properties, one has to be able to prove at least one of them. In this chapter we will discuss the most popular method for proving superconcentration. I call this "Talagrand's L^1–L^2 method".

1 Hypercontractivity

Talagrand's method is based on the use of hypercontractive inequalities. Note that by Jensen's inequality, a Markov semigroup P_t is always L^2-contractive, i.e. for all $f \in L^2(\mu)$ (where μ is the equilibrium measure) and any $t \geq 0$,

$$\|P_t f\|_{L^2(\mu)} \leq \|f\|_{L^2(\mu)}.$$

A semigroup P_t is called hypercontractive if the above contraction property can be strengthened in a certain manner: For any $p > 1$ and $t > 0$ there exists a $q = q(t, p) > p$ such that for all $f \in L^p(\mu)$

$$\|P_t f\|_{L^q(\mu)} \leq \|f\|_{L^p(\mu)}.$$

The first proof of hypercontractivity was for the Ornstein-Uhlenbeck semigroup, due to Nelson (1973) who proved that in any dimension, the OU semigroup is hypercontractive with

$$q(t, p) = 1 + (p - 1)e^{2t}.$$

This was followed by the remarkable discovery of Gross (1975) that if the associated Dirichlet form $\mathcal{E}$ satisfies a 'logarithmic Sobolev inequality' (Gross's invention), then P_t must be hypercontractive. Since then, the literature surrounding hypercontractive inequalities and their applications have seen an explosion in activity. For the proof of the hypercontractive inequality for the OU semigroup, see Appendix B.

One of the major uses of hypercontractivity is in the noise-sensitivity literature, starting with the pioneering work of Kahn et al. (1988). The so-called KKL argument was used by Talagrand (1994) to prove a quantitative version of the approximate zero-one law of Russo (1981). Talagrand's paper has an interesting variance bound, which is an improvement of a discrete Poincaré inequality. This bound was effectively used by Benjamini et al. (2003) in a breakthrough paper to prove superconcentration of the first-passage percolation time on lattices. Previously, superconcentration was known only in relatively simple models like maxima of i.i.d. random variables or in exactly computable special cases like longest increasing subsequences (Kim 1996 and Baik et al. 1999; see also Aldous and Diaconis 1999) and eigenvalues of random matrices (Tracy and Widom 1994a,b, 1996) and at least one special example from probability on trees (Dekking and Host 1991). Since the publication of Benjamini et al. (2003), similar results have been proved by various authors using various guises of the hypercontractive method (see Ledoux 2003, 2007, Aubrun 2005, Benaïm and Rossignol 2008, Benjamini and Rossignol 2008, Chatterjee 2008b, Graham 2010, Alexander and Zygouras 2012, van den Berg and Kiss 2012, Matic and Nolen 2012 and references therein).

The basic idea behind the use of hypercontractivity in proving superconcentration is the following. Suppose that g is an n-dimensional standard Gaussian random vector, and $(g^t)_{t\geq 0}$ is a stationary n-dimensional Ornstein-Uhlenbeck process with $g^0 = g$. Let P_t denote the associated semigroup and γ^n denote the n-dimensional standard Gaussian measure. Take any Borel set $A \subseteq \mathbb{R}^n$ and let $f(x) = 1_A(x)$. Take any $t > 0$ and let $p = p(t)$ solve $2 = 1 + (p-1)e^{2t}$. Then

$$\mathbb{P}(g \in A \text{ and } g^t \in A) = \mathbb{E}(f(g)f(g^t))$$
$$= \mathbb{E}_{\gamma^n}(f P_t f)$$
$$\leq \|f\|_{L^2(\gamma^n)} \|P_t f\|_{L^2(\gamma^n)}$$
$$\leq \|f\|_{L^2(\gamma^n)} \|f\|_{L^p(\gamma^n)} = (\mathbb{P}(g \in A))^{\frac{1}{2}+\frac{1}{p}}.$$

Therefore,

$$\mathbb{P}(g^t \in A \mid g \in A) \leq (\mathbb{P}(g \in A))^{\frac{1}{2}+\frac{1}{p}-1}.$$

Now note that $p < 2$, and hence the exponent on the right-hand side is strictly greater than 0. This shows that if $\mathbb{P}(g \in A)$ is very small, then the conditional probability $\mathbb{P}(g^t \in A \mid g \in A)$ drops rapidly as t increases.

In other words, even if g belongs to a given 'small set', it is likely to go out of it quickly as the vector flows over time as an OU flow.

To see why this may be connected to our interests, consider the polymer model introduced in Chap. 1, Sect. 1.2. Suppose that the probability of a certain vertex belonging to the ground state polymer (i.e. optimal path) is small. Then hypercontractivity tells us that even if the vertex belongs to the optimal path at time 0, it is likely to be dislodged from the optimal path within a small amount of time. Clearly, this should imply chaos and therefore superconcentration.

However, it is not known a priori in polymers or first-passage percolation that vertices (or edges) have small probabilities of belonging to the optimal path. It was a clever trick of Benjamini, Kalai and Schramm that allowed them to bypass this difficulty and prove the superconcentration of the passage time in first-passage percolation.

2 Talagrand's L^1–L^2 Bound

A different version of the following theorem was proved by Talagrand (1994). While Talagrand's result is applicable to functions of independent binary random variables, the version below applies to functions of Gaussian random variables.

Theorem 5.1 (Similar to Theorem 1.5 in Talagrand 1994) *Let $f : \mathbb{R}^n \to \mathbb{R}$ be an absolutely continuous function and γ^n be the n-dimensional standard Gaussian measure. Let $\partial_i f$ denote the partial derivative of f in the ith coordinate. For each i, let A_i be any number such that $\|\partial_i f\|_{L^2(\gamma^n)} \leq A_i$. Then*

$$\mathrm{Var}_{\gamma^n}(f) \leq C \sum_{i=1}^{n} \frac{A_i^2}{1 + \log(A_i/\|\partial_i f\|_{L^1(\gamma^n)})}$$

where C is a universal constant independent of f and n.

Before proving the theorem, let us try to understand what it means. Recall that the Gaussian Poincaré inequality says

$$\mathrm{Var}_{\gamma^n}(f) \leq \sum_{i=1}^{n} \|\partial_i f\|_{L^2(\gamma^n)}^2.$$

Therefore the logarithmic factor in the denominator is Talagrand's improvement over the Poincaré inequality. This gives a real improvement only when the L^2 norms of the partial derivatives of f are much larger than the L^1 norms. For example, if $f(x) = \sum_{i=1}^{n} x_i$, then $\partial_i f \equiv 1$ and so the L^1 and L^2 norms are equal. So in this case, the L^1–L^2 bound gives no improvement. (Neither should it, because the Poincaré inequality is optimal.)

On the other hand, if $f(x) = \max_{1 \leq i \leq n} x_i$, then

$$\partial_i f(x) = 1_{\{x_i \geq x_j \text{ for all } j\}}.$$

Therefore, letting $g_1, \ldots, g_n$ be i.i.d. standard Gaussian random variables, we have

$$\|\partial_i f\|_{L^2(\gamma^n)}^2 = \|\partial_i f\|_{L^1(\gamma^n)} = \mathbb{P}(g_i \geq g_j \text{ for all } j) = \frac{1}{n}.$$

So in this case the L^1–L^2 bound gives

$$\operatorname{Var}\left(\max_{1 \le i \le n} g_i\right) \le \frac{C}{\log n},$$

where C is a constant that does not depend on n. This bound is of the optimal order (see Appendix A), whereas the Poincaré inequality gives the suboptimal bound $\operatorname{Var}(\max g_i) \le C$.

However the L^1–L^2 bound does not always give the optimal bound. The following exercise gives one such instance.

Exercise 5.2 Consider a binary tree of depth n. On each edge of the tree, attach a standard Gaussian random variable and call it the 'edge-weight'. Declare the weight of a leaf to be the sum of all edge-weights of edges leading from the root to the leaf. Let M_n be the maximum of these leaf-weights. (Note that M_n is also the location of the right-most particle at the nth generation of a branching random walk (BRW) that takes Gaussian increments and splits into two offspring at each split. The expected value of M_n was proved to asymptote as $n\sqrt{2 \log 2}$ by Biggins 1977.) Prove using the L^1–L^2 bound that $\operatorname{Var}(M) \le C \log n$ for some universal constant C, improving over the bound $\operatorname{Var}(M) \le Cn$ that is given by the Poincaré inequality.

Incidentally, this bound is suboptimal. The optimal bound, as proved by Bramson and Zeitouni (2009) in a long and difficult paper, is $\operatorname{Var}(M_n) \le C$. This leads to an open question.

Open Problem 5.3 Find a 'soft proof' of the Bramson-Zeitouni result, using hypercontractivity or otherwise.

Proof of Theorem 5.1 Let P_t be the n-dimensional OU generator. From the covariance lemma and our previous observation that

$$\nabla P_t = e^{-t} P_t \nabla,$$

it follows that

$$\operatorname{Var}_{\gamma^n}(f) = \int_0^\infty e^{-t} \sum_{i=1}^n \mathbb{E}_{\gamma^n}(\partial_i f \, P_t \partial_i f) \, dt.$$

Now fix some $t > 0$ and take any $h \in L^2(\gamma^n)$. Let $p = 1 + e^{-2t}$, so that $2 = 1 + e^{2t}(p-1)$. Then by the Cauchy-Schwarz inequality and the hypercontractive property of the OU semigroup,

$$\mathbb{E}_{\gamma^n}(h P_t h) \le \|h\|_{L^2(\gamma^n)} \|P_t h\|_{L^2(\gamma^n)}$$

$$\le \|h\|_{L^2(\gamma^n)} \|h\|_{L^p(\gamma^n)}.$$

Since $1 < p < 2$, writing $|h|^p = |h|^{2p-2}|h|^{2-p}$ and applying Hölder's inequality gives

$$\|h\|_{L^p(\gamma^n)} \le \|h\|_{L^2(\gamma^n)}^{2-\frac{2}{p}} \|h\|_{L^1(\gamma^n)}^{\frac{2}{p}-1}.$$

Thus, if A is a number such that $\|h\|_{L^2(\gamma^n)} \le A$, then combining the last two displays we have

$$\mathbb{E}_{\gamma^n}(h P_t h) \le \|h\|_{L^2(\gamma^n)}^{3-\frac{2}{p}} \|h\|_{L^1(\gamma^n)}^{\frac{2}{p}-1}$$

$$\le A^{3-\frac{2}{p}} \|h\|_{L^1(\gamma^n)}^{\frac{2}{p}-1} = A^2 \left(\frac{\|h\|_{L^1(\gamma^n)}}{A}\right)^{\tanh t}.$$

Therefore, writing

$$a_i := \frac{\|\partial_i f\|_{L^1(\gamma^n)}}{A_i},$$

and applying the inequality $\tanh t \ge 1 - e^{-t}$, we get

$$\mathrm{Var}_{\gamma^n}(f) \le \sum_{i=1}^n A_i^2 \int_0^\infty e^{-t} a_i^{1-e^{-t}} dt$$

$$= \sum_{i=1}^n A_i^2 \int_0^1 a_i^u du \le C \sum_{i=1}^n \frac{A_i^2}{1+\log(1/a_i)},$$

where C is a universal constant. $\square$

3 Talagrand's Method Always Works for Monotone Functions

The L^1–L^2 method is a powerful tool for proving superconcentration, but does it always work? In this section we will see that the method is guaranteed to work for monotone functions. In other words, if a monotone function of a collection of Gaussian random variables is superconcentrated, then Talagrand's L^1–L^2 method can be used to prove it. On the other hand, we will see in the next chapter that the method may not work for functions that are not monotone.

Theorem 5.4 *Let $f : \mathbb{R}^n \to \mathbb{R}$ be an absolutely continuous function that is monotone in each coordinate. Let all notation be as in Sect. 2. Define*

$$a := \frac{\mathrm{Var}_{\gamma^n}(f)}{\mathbb{E}_{\gamma^n}|\nabla f|^2} \quad \text{and} \quad b := \frac{\sum_{i=1}^n \|\partial_i f\|_{L^1(\gamma^n)}^2}{\sum_{i=1}^n \|\partial_i f\|_{L^2(\gamma^n)}^2}.$$

Then a and b lie between 0 and 1 and satisfy the inequalities

$$b \le a \le \frac{C}{1 + \log(b^{-1/2})},$$

where C is a universal constant. This implies that a is small if and only if b is small.

Proof For each i, let $u_i := \|\partial_i f\|_{L^1(\gamma^n)}^2$ and $v_i := \|\partial_i f\|_{L^2(\gamma^n)}^2$. Define

$$g(x) := \frac{1}{1 + \log(x^{-1/2})} = -\frac{2}{\log(x/e^2)}.$$

It is easily verified (by computing the second derivative) that the map $y \mapsto 1/\log y$ is convex in the interval $(0, e^{-2})$. From this it follows that g is concave on $(0, 1)$. Therefore Theorem 5.1 and Jensen's inequality give

$$\mathrm{Var}_{\gamma^n}(f) \le C \sum_{i=1}^n \frac{\|\partial_i f\|_{L^2(\gamma^n)}^2}{1 + \log(\|\partial_i f\|_{L^2(\gamma^n)}/\|\partial_i f\|_{L^1(\gamma^n)})}$$

$$= C \sum_{i=1}^n g(u_i/v_i) v_i$$

$$\le C\, g\!\left(\frac{\sum_{i=1}^n (u_i/v_i) v_i}{\sum_{i=1}^n v_i}\right) \sum_{i=1}^n v_i = C\, g(b)\, \mathbb{E}_{\gamma^n} |\nabla f|^2.$$

This proves the upper bound. For the lower bound, let $c_i := \mathbb{E}_{\gamma^n}(\partial_i f)$, $i = 1, \ldots, n$. Define

$$h(x) := f(x) - \sum_{i=1}^n c_i x_i.$$

Then note that for each i,

$$\mathrm{Cov}_{\gamma^n}(h, x_i) = \mathrm{Cov}_{\gamma^n}(f, x_i) - c_i,$$

since $\mathrm{Cov}_{\gamma^n}(x_i, x_j) = 0$ if $i \ne j$ and $= 1$ if $i = j$. However, Gaussian integration by parts (see Appendix A) gives

$$\mathrm{Cov}_{\gamma^n}(f, x_i) = \mathbb{E}_{\gamma^n}(x_i f) = \mathbb{E}_{\gamma^n}(\partial_i f) = c_i.$$

Thus, $\mathrm{Cov}_{\gamma^n}(h, x_i) = 0$ for each i. Therefore

$$\mathrm{Var}_{\gamma^n}(f) = \mathrm{Var}_{\gamma^n}\!\left(h + \sum_{i=1}^n c_i x_i\right) = \mathrm{Var}_{\gamma^n}(h) + \sum_{i=1}^n c_i^2 \ge \sum_{i=1}^n c_i^2.$$

Finally note that by the monotonicity of f, $c_i = \|\partial_i f\|_{L^1(\gamma^n)}$. This completes the proof of the lower bound. □

Theorem 5.4 has an interesting consequence for a generalized polymer (or last-passage percolation) model. Consider an n-dimensional standard Gaussian random vector $g = (g_1, \ldots, g_n)$, and a function $f : \mathbb{R}^n \to \mathbb{R}$ defined as

$$f(x) = \max_{p \in \mathcal{P}} \sum_{i \in p} x_i,$$

where $\mathcal{P}$ is a collection of subsets of $\{1, \ldots, n\}$ that may be called 'paths' in a generalized sense. This is a broad generalization of the lattice polymer model that was introduced in Chap. 1, Sect. 1.2. Clearly, f is monotone in each coordinate. Let $\hat{p}$ denote the member of $\mathcal{P}$ that maximizes $\sum_{i \in p} g_i$. Then it is easy to see that $\hat{p}$ is well-defined almost surely and

$$\|\partial_i f\|_{L^1(\gamma^n)} = p_i \quad \text{and} \quad \|\partial_i f\|_{L^2(\gamma^n)} = p_i^{1/2},$$

where $p_i = \mathbb{P}(i \in \hat{p})$. Let $g' = (g'_1, \ldots, g'_n)$ be another standard Gaussian random vector, independent of g. Let $\hat{p}'$ be the optimal element of $\mathcal{P}$ in this new environment. Note that

$$\sum_{i=1}^n p_i = \mathbb{E}|\hat{p}| \quad \text{and} \quad \sum_{i=1}^n p_i^2 = \mathbb{E}|\hat{p} \cap \hat{p}'|,$$

where $|\cdot|$ denotes the cardinality of a set. Therefore, Theorem 5.4 implies that $f(g)$ is superconcentrated if and only if $\mathbb{E}|\hat{p} \cap \hat{p}'|$ is very small compared to $\mathbb{E}|\hat{p}|$. That is, superconcentration holds if and only if the expected number of vertices in the optimal path is much larger than the expected size of the overlap between the optimal paths from two independent environments.

4 The Benjamini-Kalai-Schramm Trick

Recall the first-passage percolation model on $\mathbb{Z}^d$ with i.i.d. non-negative edge weights. Let T_n be the first-passage time from the origin to the point ne_1, where $e_1 = (1, 0, 0, \ldots, 0)$. Under mild conditions on the edge weights, Kesten (1993) proved that $\text{Var}(T_n) \leq Cn$, where C depends only on the distribution of the edge weights and the dimension d. This was improved by Benjamini et al. (2003), who proved that for binary edge weights in dimension ≥ 2, $\text{Var}(T_n) \leq Cn/\log n$. This result was later extended for a larger class of edge-weights by Benaïm and Rossignol (2008). The basic tool in Benjamini et al. (2003) is the L^1–L^2 bound; but the L^1–L^2 bound alone does not suffice due to a certain technical difficulty that we will see below. There is a nice little trick in Benjamini et al. (2003) that allowed the authors to circumvent this problem. I call this the 'BKS trick'.

Let us now investigate the difficulty in directly applying the L^1–L^2 inequality to first-passage percolation. We assume that the edge-weight distribution is a *Lipschitz function of the Gaussian distribution, and is bounded away from zero and infinity.* In other words, if ω_e is the weight of an edge, assume that there is a standard Gaussian random variables g_e and an absolutely continuous function $F : \mathbb{R} \to \mathbb{R}$ such that $|F'|$ is uniformly bounded by a constant K and there are constants $0 < a < b < \infty$ such that $a \leq F(x) \leq b$ for all x. The uniform distribution on an interval $[a, b]$ is an example of such a distribution.

The assumption that the edge weights are uniformly bounded away from zero and infinity implies that the length of the optimal path between any two vertices cannot exceed a multiple of their Euclidean distance. This implies, in particular, that T_n is a function of only finitely many edge weights.

Let $\hat{p}_n$ be the optimal path from the origin to ne_1. With obvious abuse of notation, this gives

$$\frac{\partial T_n}{\partial g_e} = F'(g_e) 1_{\{e \in \hat{p}_n\}}.$$

Consequently,

$$\left\| \frac{\partial T_n}{\partial g_e} \right\|_{L^1} \leq K p_e,$$

where $p_e := \mathbb{P}(e \in \hat{p}_n)$. Therefore, to apply the L^1–L^2 bound we certainly need to show that the p_e's are small for 'most' edges. Unfortunately this is not something that has any direct proof. (One can prove such a statement *after* proving superconcentration of the first-passage time.)

We are now ready to prove Theorem 1.1 of Chap. 1.

Proof of Theorem 1.1 of Chap. 1 Let F be a Lipschitz function with $|F'| \leq K$, such that the weight of edge e is $F(g_e)$, where g_e are i.i.d. standard Gaussian random variables.

Take any n and fix k, to be chosen later. Let $T = T_n$. Let B be a box of side length k centered at the origin. Let $|B|$ denote the number of lattice points in B. For each lattice point $x \in B$, let T_x be the first-passage time from x to $x + ne_1$. Let

$$\tilde{T} := \frac{1}{|B|} \sum_{x \in B} T_x.$$

Since the edge-weights are uniformly bounded away from zero and infinity, it is easy to see that $|T_x - T| \leq Ck$ for each $x \in B$, where C is a constant depending only on the distribution of the edge-weights and the dimension. (Henceforth, C will denote any such constant.) Consequently, $|\tilde{T} - T| \leq Ck$, and therefore

$$\text{Var}(T) \leq 2\,\text{Var}(\tilde{T}) + Ck^2. \tag{5.1}$$

Let $\hat{p}$ denote the optimal path from the origin to ne_1. For each $x \in B$, let $\hat{p}_x$ denote the optimal path from x to $ne_1 + x$. Then for any edge e,

$$\frac{\partial \tilde{T}}{\partial g_e} = \frac{F'(g_e)}{|B|} \sum_{x \in B} 1_{\{e \in \hat{p}_x\}}.$$

4 The Benjamini-Kalai-Schramm Trick

Consequently,
$$\left\|\frac{\partial \widetilde{T}}{\partial g_e}\right\|_{L^1} \le \frac{K}{|B|} \sum_{x \in B} \mathbb{P}(e \in \hat{p}_x).$$

Let $e - x$ denote the edge e 'translated by x'. Then by translation invariance of the problem,
$$\mathbb{P}(e \in \hat{p}_x) = \mathbb{P}(e - x \in \hat{p}).$$

Let $e - B$ denote the box B 'translated by e', that is, the set of all edges $e - x$ as x runs over vertices of B. The above two displays imply that
$$\left\|\frac{\partial \widetilde{T}}{\partial g_e}\right\|_{L^1} \le \frac{K}{|B|} \mathbb{E}\left|(e - B) \cap \hat{p}\right| =: A'_e. \tag{5.2}$$

Now $|B|$ is of the order k^d, whereas the number of edges in $e - B$ that belong to $\hat{p}$ cannot exceed a constant multiple of k because $(e - B) \cap \hat{p}$ is a subset of the minimum weight path between the first entry point of $\hat{p}$ in $e - B$ and the last exit point, and these two points are at distance less than a multiple of k. Therefore
$$A'_e \le C k^{1-d}. \tag{5.3}$$

On the other hand
$$\left\|\frac{\partial \widetilde{T}}{\partial g_e}\right\|_{L^2}^2 \le \mathbb{E}\left(\frac{K}{|B|} \sum_{x \in B} 1_{\{e \in \hat{p}_x\}}\right)^2$$
$$\le \mathbb{E}\left(\frac{K^2}{|B|} \sum_{x \in B} 1^2_{\{e \in \hat{p}_x\}}\right) = \frac{K^2}{|B|} \sum_{x \in B} \mathbb{P}(e \in \hat{p}_x)$$
$$= \frac{K^2}{|B|} \sum_{x \in B} \mathbb{P}(e - x \in \hat{p}) = \frac{K^2}{|B|} \mathbb{E}\left|(e - B) \cap \hat{p}\right| =: A_e^2.$$

Note that $A_e^2 = K A'_e$, and therefore by (5.2) and (5.3),
$$\left\|\frac{\partial \widetilde{T}}{\partial g_e}\right\|_{L^1} \le A'_e = \sqrt{A'_e} \frac{A_e}{\sqrt{K}} \le C k^{-\frac{d-1}{2}} A_e.$$

Therefore by the L^1–L^2 inequality,
$$\operatorname{Var}(\widetilde{T}) \le \frac{C}{\log k} \sum_e A_e^2$$
$$\le \frac{C}{\log k} \frac{1}{|B|} \sum_{x \in B} \sum_e \mathbb{P}(e - x \in \hat{p})$$

$$= \frac{C}{\log k} \mathbb{E}|\hat{p}| \leq \frac{Cn}{\log k}.$$

Combining this with (5.1) we get

$$\mathrm{Var}(T) \leq \frac{Cn}{\log k} + Ck^2.$$

The proof is completed by choosing $k = n^\alpha$ for any $\alpha < 1/2$. □

5 Superconcentration in Gaussian Polymers

The proof of superconcentration of the ground state energy in the $(1+1)$-dimensional Gaussian polymer model follows exactly along the lines of the BKS proof in first-passage percolation, with an added technical difficulty in controlling the size of $T - T_x$. The reader has to think about the problem to appreciate the nature of this difficulty; it is hard to explain otherwise. The superconcentration result for the $(1+1)$-dimensional polymer, which is equivalent to 2-dimensional last-passage percolation, was proved in Chatterjee (2008b). The result was extended to $(d+1)$-dimensions in Graham (2010). The said technical difficulty in considerably harder in $d \geq 2$. Here we reproduce a simplified version of the proof in $(1+1)$-dimension from Chatterjee (2008b).

Proof of Theorem 1.3 of Chap. 1 Throughout this proof, the constant C denotes any positive universal constant.

Fix n. Also fix $k = k(n)$, to be chosen later depending on n. Let B be the set of lattice points in $\{0\} \times \mathbb{Z}$ that are at distance $\leq k$ from the origin, and all of whose coordinates are even. Let $|B|$ denote the number of points in B. For each lattice point $x \in B$, let E_n^x be the minimum energy among all polymer paths of length n that originate at x. Let

$$\widetilde{E}_n := \frac{1}{|B|} \sum_{x \in B} E_n^x.$$

Let $\hat{p}$ be the optimal (i.e. minimum energy) path of length n starting at the origin. Let $(0, v_0), (1, v_1), \ldots, (n, v_n)$ be the vertices of $\hat{p}$, where $v_0 = 0$ and $|v_{j+1} - v_j| = 1$ for each j. Note that

$$E_{n+1} \leq E_n - \max\{g_{(n+1, v_n+1)}, g_{(n+1, v_n-1)}\},$$

which shows that there is a positive constant C such that $E_{n+1} - E_n \leq -C$. Consequently, for any $0 \leq m \leq n < \infty$,

$$\mathbb{E}(E_n) - \mathbb{E}(E_m) \leq -C(n-m). \tag{5.4}$$

Let τ be the minimum j such that $v_{j+1} - v_j = 1$. If there is no such j, let $\tau = n$. Take any $m \leq n$. Let $E_{n,m}$ be the minimum energy among all polymer

5 Superconcentration in Gaussian Polymers

paths of length n, originating at the origin and passing through $(m, -m)$. Then clearly $\mathbb{E}(E_{n,m}) = \mathbb{E}(E_{n-m})$, since the first m vertices in any such path are deterministic and the expected value of their weights is zero. Therefore by (5.4), $\mathbb{E}(E_n - E_{n,m}) \le -Cm$. From this, it may be easily proved using the Gaussian concentration inequality ((A.5) in Appendix A) that

$$\mathbb{P}(E_n = E_{n,m}) \le \mathbb{P}(E_n - E_{n,m} \ge 0) \le e^{-Cm^2/n}.$$

Now note that the event $\{\tau \ge m\}$ implies that $E_n = E_{n,m}$. Thus,

$$\mathbb{P}(\tau \ge m) \le e^{-Cm^2/n}. \tag{5.5}$$

Let x be the point $(0, 2)$. Note that the sequence of points $(0,2), (1, 2+v_1), (2, 2+v_2), \ldots, (\tau, 2+v_\tau), (\tau+1, v_{\tau+1}), (\tau+2, v_{\tau+2}), \ldots, (n, v_n)$ is a legitimate polymer path of length n starting at x. Also note that $v_j = -j$ for $j \le \tau$. Consequently,

$$E_n^x \le -\sum_{j=0}^{\tau} g(j, 2-j) - \sum_{j=\tau+1}^{n} g(j, v_j)$$

$$= E_n - \sum_{j=0}^{\tau} g(j, 2-j) + \sum_{j=0}^{\tau} g(j, -j). \tag{5.6}$$

Lastly, another easy application of Gaussian concentration inequality (A.5) from Appendix A shows that

$$\mathbb{E} \max_{0 \le m \le n} \left| \frac{\sum_{j=0}^{m} g(j, -j)}{\sqrt{m+1}} \right|^2 \le C \log n, \tag{5.7}$$

and the same bound holds if $g(j, -j)$ is replaced by $g(j, 2-j)$.

Combining (5.5), (5.6) and (5.7), it is now easy to deduce that

$$\mathbb{E}(E_n^x - E_n)_+^2 \le C\sqrt{n} \log n,$$

where $(E_n^x - E_n)_+$ is the positive part of $E_n^x - E_n$. By symmetry, the same bound holds for the negative part as well. Thus, for $x = (0, 2)$,

$$\mathbb{E}(E_n^x - E_n)^2 \le C\sqrt{n} \log n.$$

Now take $x = (0, 2l)$ for some positive integer l. Then by the above inequality, the Cauchy-Schwarz inequality and translation symmetry,

$$\mathbb{E}(E_n^x - E_n)^2 \le l \sum_{j=1}^{l} \mathbb{E}(E_n^{(0,2j)} - E_n^{(0,2j-2)})^2 \le Cl\sqrt{n} \log n.$$

Clearly the same bound holds if l is negative, with l replaced by $|l|$ on the right. Thus, for any $x \in B$,
$$\mathbb{E}(E_n^x - E_n)^2 \leq Ck\sqrt{n}\log n,$$
and therefore
$$\mathrm{Var}(E_n) \leq 2\ \mathrm{Var}(\widetilde{E}_n) + 2\ \mathbb{E}(\widetilde{E}_n - E_n)^2 \leq 2\ \mathrm{Var}(\widetilde{E}_n) + Ck\sqrt{n}\log n.$$

Proceeding exactly as in the proof of Theorem 1.1, one can now show that $\mathrm{Var}(\widetilde{E}_n) \leq Cn/\log k$. (We leave the details as an exercise for the reader.) Choosing $k = n^\alpha$ for any $\alpha < 1/2$ finishes the proof. □

Exercise 5.5 Combining Theorems 4.1, 1.3 and 3.5, give a proof of Theorem 1.17.

6 Sharpness of the Logarithmic Improvement

The improvement factor of $\log n$ is possibly sub-optimal in most cases, but there are cases where it may actually give the correct answer. A simple instance is given by the maximum of independent Gaussian random variables, where the L^1–L^2 method is capable of producing the right answer. But one may argue that this case is rather special and devoid of interesting correlations.

The following example, repeated from Sect. 1.2 of Chap. 1, is a simple modification of the polymer model that demonstrates the sharpness of the L^1–L^2 bound even in cases involving non-trivial correlations.

Let $(g_{ij})_{1 \leq i,j \leq n}$ be a square array of standard Gaussian random variables. Let Σ be the set of all maps from $\{1, \ldots, n\}$ into itself. For each $\sigma \in \Sigma$, let
$$f(\sigma) := \sum_{i=1}^{n} g_{i\sigma(i)}.$$

Then f is a Gaussian field, and
$$\mathrm{Cov}(f(\sigma), f(\sigma')) = |\{i : \sigma_i = \sigma_i'\}|.$$

Note that f is a 'superfield' of the field of all paths of length n in a $(1+1)$-dimensional Gaussian random polymer. Now, clearly,
$$\max_\sigma f(\sigma) = \sum_{i=1}^n \max_{1 \leq j \leq n} g_{ij}.$$

Therefore the variance of the maximum is of order $n/\log n$, since each term on the right-hand side in the above display has variance of order $1/\log n$ (being a maximum of n independent Gaussian random variables), and the terms are independent. The L^1–L^2 method gives the correct order of the variance of the maximum in this example.

Chapter 6
The Spectral Method for Proving Superconcentration

The primary goal of this chapter is to prove the superconcentration of the free energy of the SK model, defined in Chap. 1, Sect. 1.3. It turns out, as we shall see below, that the L^1–L^2 method does not work for this problem. We will take a different route, that may be called the 'spectral approach' for proving superconcentration.

First let us see why the L^1–L^2 method does not work for the SK model. Recall that the free energy at inverse temperature β is defined as

$$F_n(\beta) = -\frac{1}{\beta} \log \sum_{\sigma \in \{-1,1\}^n} e^{-\beta H_n(\sigma)},$$

where H_n is the Hamiltonian

$$H_n(\sigma) = -\frac{1}{\sqrt{n}} \sum_{1 \le i < j \le n} g_{ij} \sigma_i \sigma_j,$$

with g_{ij} being i.i.d. standard Gaussian random variables. We will now fix β and denote $F_n(\beta)$ by F_n.

For a function $f : \{-1, 1\}^n \to \mathbb{R}$, define

$$\langle f(\sigma) \rangle := \frac{\sum_\sigma f(\sigma) e^{-\beta H_n(\sigma)}}{\sum_\sigma e^{-\beta H_n(\sigma)}}.$$

This is nothing but the conditional expectation of $f(\sigma)$ given the disorder $g = (g_{ij})_{1 \le i < j \le n}$, where σ is picked from the Gibbs measure at inverse temperature β. With this notation, it is easy to see that

$$\frac{\partial F_n}{\partial g_{ij}} = -\frac{1}{\sqrt{n}} \langle \sigma_i \sigma_j \rangle.$$

The random variable $\langle \sigma_i \sigma_j \rangle$ is bounded between -1 and 1. Moreover, its distribution does not depend on i and j. The following exercise shows that the L^1–L^2 method is ineffective for proving superconcentration of the free energy at large enough β.

S. Chatterjee, *Superconcentration and Related Topics*,
Springer Monographs in Mathematics, DOI 10.1007/978-3-319-03886-5_6,
© Springer International Publishing Switzerland 2014

Exercise 6.1 Prove that when β is large enough, both the L^1 and L^2 norms of $\langle \sigma_i \sigma_j \rangle$ remain uniformly bounded away from zero as $n \to \infty$.

Note that, as proved in Sect. 3 of Chap. 5, the L^1–L^2 method can always be used to prove superconcentration for monotone functions. Unfortunately the free energy is not a monotone function of the disorder.

1 Spectral Decomposition of the OU Semigroup

We begin with a discussion of Hermite polynomials. For each $k \geq 0$, let

$$H_k(x) := (-1)^k e^{x^2/2} \frac{d^k}{dx^k} e^{-x^2/2}.$$

For example, $H_0 \equiv 1$, $H_1(x) = x$, $H_2(x) = x^2 - 1$, and so on. These are known as the Hermite polynomials. It is a well-known fact that the Hermite polynomials form an orthogonal basis of $L^2(\gamma)$, where γ is the one-dimensional standard Gaussian measure. Moreover, if L is the OU generator $Lf(x) = f''(x) - xf'(x)$, then H_k is an eigenfunction of $-L$ with eigenvalue k. Moreover, these eigenvalues and eigenfunctions give the complete spectral decomposition of the one-dimensional OU generator.

The multidimensional case is similar. Let L be the n-dimensional OU generator, i.e. $Lf(x) = \Delta f(x) - x \cdot \nabla f(x)$. The eigenvalues and eigenfunctions of $-L$ are now indexed by $\mathbb{Z}_+^n$, where $\mathbb{Z}_+ = \{0, 1, 2, \ldots\}$. For each $k = (k_1, \ldots, k_n) \in \mathbb{Z}_+^n$, the kth multidimensional Hermite polynomial H_k (not to be confused with any Hamiltonian H_n) is defined as

$$H_k(x) = \prod_{i=1}^n H_{k_i}(x_i).$$

Here x stands for the vector $(x_1, \ldots, x_n)$. The function H_k is an eigenfunction of $-L$ corresponding to the eigenvalue $k_1 + \cdots + k_n$.

The one-dimensional Hermite polynomial H_k satisfies $\|H_k\|_{L^2(\gamma)}^2 = k!$. Consequently, the n-dimensional Hermite polynomial H_k satisfies

$$\|H_k\|_{L^2(\gamma^n)}^2 = k! := k_1! k_2! \cdots k_n!.$$

Therefore by the Plancherel formula, we have that for any $f \in L^2(\gamma^n)$,

$$\|f\|_{L^2(\gamma^n)}^2 = \sum_{k \in \mathbb{Z}_+^n} \frac{(f, H_k)^2}{k!}.$$

1 Spectral Decomposition of the OU Semigroup

If f is a smooth function, there is a nice way to compute the inner products with the Hermite polynomials. First, consider the one-dimensional case. Applying integration by parts, it follows easily that

$$(f, H_k) = \frac{1}{\sqrt{2\pi}} \int_{-\infty}^{\infty} f(x)(-1)^k \frac{d^k}{dx^k} e^{-x^2/2} dx = (f', H_{k-1}).$$

Continuing, we arrive at the formula

$$(f, H_k) = \left(f^{(k)}, H_0\right) = \mathbb{E}_\gamma\left(f^{(k)}\right),$$

where $f^{(k)}$ is the kth derivative of f. This formula generalizes to n dimensions in a straightforward manner: For $k = (k_1, \ldots, k_n)$ and any smooth f with all derivatives in $L^2(\gamma^n)$,

$$(f, H_k) = \mathbb{E}_{\gamma^n}\left(\frac{\partial^{k_1+\cdots+k_n} f}{\partial x_1^{k_1} \cdots \partial x_n^{k_n}}\right).$$

Therefore if f is a smooth function with all derivatives in $L^2(\gamma^n)$, then

$$\|f\|_{L^2(\gamma^n)}^2 = \sum_{k \in \mathbb{Z}_+^n} \frac{1}{k!} \left(\mathbb{E}_{\gamma^n}\left(\frac{\partial^{k_1+\cdots+k_n} f}{\partial x_1^{k_1} \cdots \partial x_n^{k_n}}\right)\right)^2.$$

The sum can be reorganized as

$$\|f\|_{L^2(\gamma^n)}^2 = \sum_{m=0}^{\infty} \frac{1}{m!} \sum_{k \in \mathbb{Z}_+^n,\, k_1+\cdots+k_n=m} \frac{m!}{k!} \left(\mathbb{E}_{\gamma^n}\left(\frac{\partial^m f}{\partial x_1^{k_1} \cdots \partial x_n^{k_n}}\right)\right)^2$$

$$= \sum_{m=0}^{\infty} \frac{\theta_m(f)}{m!},$$

where

$$\theta_m(f) := \sum_{1 \le i_1,\ldots,i_m \le n} \left(\mathbb{E}_{\gamma^n}\left(\frac{\partial^m f}{\partial x_{i_1} \cdots \partial x_{i_m}}\right)\right)^2. \tag{6.1}$$

An alternative way to write this is:

$$\mathrm{Var}_{\gamma^n}(f) = \sum_{m=1}^{\infty} \frac{\theta_m(f)}{m!}. \tag{6.2}$$

One may attempt to apply this identity to compute the variance of the free energy of the SK model and in particular to prove that the variance may be bounded by a constant independent of n. Indeed, the results are encouraging since (as we shall see later) one can prove that for each m, $\theta_m(F_n)$ may be bounded by a constant $C_m = C_m(\beta)$ independent of n. Unfortunately I could not get the C_m's to satisfy

$\sum_{m=1}^{\infty} C_m/m! < \infty$. Still, there is a way to prove superconcentration even with crude bounds for $\theta_m(f)$. This is the topic of the next section.

2 An Improved Poincaré Inequality

Recall our usual setting: P_t is a reversible Markov semigroup, with generator L, Dirichlet form $\mathcal{E}$, equilibrium measure μ, eigenvalues $(\lambda_k)_{k \geq 0}$ arranged in increasing order, and eigenfunctions $(u_k)_{k \geq 0}$. The following result improves the usual Poincaré inequality.

Theorem 6.2 *Take any $m \geq 1$ and suppose that $\lambda_m > 0$. Then for any function $f \in L^2(\mu)$,*

$$\mathrm{Var}_\mu(f) \leq \sum_{k=1}^{m-1} (u_k, f)^2 + \frac{\mathcal{E}(f, f)}{\lambda_m}.$$

Proof Note that

$$\mathrm{Var}_\mu(f) = \sum_{k=1}^{m-1} (u_k, f)^2 + \sum_{k=m}^{\infty} (u_k, f)^2$$

$$\leq \sum_{k=1}^{m-1} (u_k, f)^2 + \frac{1}{\lambda_m} \sum_{k=m}^{\infty} \lambda_k (u_k, f)^2$$

$$\leq \sum_{k=1}^{m-1} (u_k, f)^2 + \frac{1}{\lambda_m} \sum_{k=1}^{\infty} \lambda_k (u_k, f)^2 = \sum_{k=1}^{m-1} (u_k, f)^2 + \frac{\mathcal{E}(f, f)}{\lambda_m}.$$

This completes the proof of the theorem. □

3 Superconcentration in the SK Model

For the OU semigroup, Theorem 6.2 implies that for any $m \geq 1$,

$$\mathrm{Var}_{\gamma^n}(f) \leq \sum_{k=1}^{m-1} \frac{\theta_k(f)}{k!} + \frac{\mathbb{E}_{\gamma^n} |\nabla f|^2}{m}, \qquad (6.3)$$

where $\theta_k(f)$ is defined in (6.1). This is used to prove the following theorem.

Theorem 6.3 *Let $F_n(\beta)$ be the free energy of the SK model at inverse temperature β. Then*

$$\mathrm{Var}(F_n(\beta)) \leq \frac{C(\beta) n \log \log n}{\log n},$$

3 Superconcentration in the SK Model

where $C(\beta)$ is a constant that depends only on β.

Note that this is a slightly weaker version of Theorem 1.6 of Chap. 1. We will prove Theorem 1.6 in Chap. 10.

Proof Let $\langle \cdot \rangle$ denote averaging under the Gibbs measure at inverse temperature β. Let $\mathbb{E}$ denote the unconditional averaging, i.e. averaging over the Gibbs measure as well as the disorder.

Let $\sigma^1, \ldots, \sigma^m$ denote m i.i.d. configurations drawn from the Gibbs measure. Note that $\sigma^1, \ldots, \sigma^m$ are conditionally independent given the disorder g, but unconditionally dependent. Moreover, by the symmetry of the problem, the unconditional distribution of each σ^ℓ is the uniform distribution on the hypercube $\{-1, 1\}^n$.

Fix β and let $F_n = F_n(\beta)$. We claim that for any $i_1, j_1, i_2, j_2, \ldots, i_k, j_k$,

$$\frac{\partial^k F_n}{\partial g_{i_1 j_1} \cdots \partial g_{i_k j_k}} = \frac{\beta^{k-1}}{n^{k/2}} \sum_{1 \le \ell_1, \ldots, \ell_k \le k} c_k(\ell_1, \ldots, \ell_k) \langle \sigma_{i_1}^{\ell_1} \sigma_{j_1}^{\ell_1} \cdots \sigma_{i_k}^{\ell_k} \sigma_{j_k}^{\ell_k} \rangle, \quad (6.4)$$

where $c_k(\ell_1, \ldots, \ell_k)$ are constants such that for all $1 \le \ell_1, \ldots, \ell_k \le k$,

$$|c_k(\ell_1, \ldots, \ell_k)| \le (k-1)!. \quad (6.5)$$

This is easy to prove by induction on k. It is true for $k = 1$ since

$$\frac{\partial F_n}{\partial g_{ij}} = -\frac{1}{\sqrt{n}} \langle \sigma_i \sigma_j \rangle.$$

(This also shows that $|\nabla F|^2 \le n$ everywhere.)

Suppose that it is true for some k. An easy computation shows that one can write

$$\frac{\partial}{\partial g_{i_{k+1} j_{k+1}}} \langle \sigma_{i_1}^{\ell_1} \sigma_{j_1}^{\ell_1} \cdots \sigma_{i_k}^{\ell_k} \sigma_{j_k}^{\ell_k} \rangle$$

$$= \frac{\beta}{\sqrt{n}} \sum_{\ell_{k+1}=1}^{k+1} d_{k+1}(\ell_1, \ldots, \ell_{k+1}) \langle \sigma_{i_1}^{\ell_1} \sigma_{j_1}^{\ell_1} \cdots \sigma_{i_{k+1}}^{\ell_{k+1}} \sigma_{j_{k+1}}^{\ell_{k+1}} \rangle,$$

where $|d_{k+1}(\ell_1, \ldots, \ell_k, \ell)| \le 1$ for all $1 \le \ell \le k$ and $|d_{k+1}(\ell_1, \ldots, \ell_k, k+1)| \le k$. It is now easy to complete the inductive step, which proves (6.4).

Consider an independent system with disorder $\tilde{g} = (\tilde{g}_{ij})_{1 \le i < j \le n}$, that is again a collection of independent standard Gaussian random variables, independent of the g_{ij}'s. Let $\tilde{\sigma}^1, \ldots, \tilde{\sigma}^k$ be i.i.d. configurations drawn according to the Gibbs measure induced by the disorder $\tilde{g}$. Then note that

$$\left(\mathbb{E}\big(\sigma_{i_1}^{\ell_1} \sigma_{j_1}^{\ell_1} \cdots \sigma_{i_k}^{\ell_k} \sigma_{j_k}^{\ell_k} \big) \right)^2 = \mathbb{E}\big(\sigma_{i_1}^{\ell_1} \tilde{\sigma}_{i_1}^{\ell_1} \sigma_{j_1}^{\ell_1} \tilde{\sigma}_{j_1}^{\ell_1} \cdots \sigma_{i_k}^{\ell_k} \tilde{\sigma}_{i_k}^{\ell_k} \sigma_{j_k}^{\ell_k} \tilde{\sigma}_{j_k}^{\ell_k} \big).$$

This identity, (6.4) and the bound (6.5) show that

$$\sum_{1\leq i_1,j_1,\ldots,i_k,j_k\leq n} \left(\mathbb{E}\frac{\partial^k F_n}{\partial g_{i_1 j_1}\cdots \partial g_{i_k j_k}}\right)^2$$

$$\leq \frac{\beta^{2k-2}(k-1)!}{n^k} \sum_{1\leq \ell_1,\ldots,\ell_k\leq k} \mathbb{E}\big((\sigma^{\ell_1}\cdot\tilde{\sigma}^{\ell_1})^2\cdots(\sigma^{\ell_k}\cdot\tilde{\sigma}^{\ell_k})^2\big)$$

$$\leq \frac{\beta^{2k-2}(k-1)!k^k}{n^k}\mathbb{E}\big((\sigma^1\cdot\tilde{\sigma}^1)^{2k}\big),$$

where the last step follows from Hölder's inequality and symmetry among replicas.

Since σ^1 and $\tilde{\sigma}^1$ are unconditionally independent and uniformly distributed on the hypercube, $\sigma^1\cdot\tilde{\sigma}^1$ is simply the sum of n i.i.d. symmetric $\{-1, 1\}$-valued random variables. Therefore by the well known inequality of Hoeffding (1963), it follows easily that

$$\mathbb{E}\big((\sigma^1\cdot\tilde{\sigma}^1)^{2k}\big) \leq k^k n^k.$$

Thus, for each $k \geq 1$,

$$\theta_k(F) \leq \beta^{2k-2} k^{3k}.$$

Using this and the previously observed bound $|\nabla F|^2 \leq n$, the inequality (6.3) finishes the proof upon taking $m = c\log n/\log\log n$ for some sufficiently small absolute constant c. □

Exercise 6.4 Combining Theorems 4.2, 6.3 and 3.5, give a proof of Theorem 1.16.

Chapter 7
Independent Flips

Let μ^n be a product probability measure on a product space $\mathcal{X}^n$, where μ is a probability measure on a measurable space $\mathcal{X}$. There is a canonical Markov process that has μ^n as equilibrium measure. I call this the 'independent flips' process. It is described quite simply: Take n independent standard Poisson clocks. (A standard Poisson clock is a clock that rings at times $T_1, T_2, \ldots$, where $\{T_i\}_{i=1,2,\ldots}$ is a Poisson process with rate 1.) Define a Markov process on $\mathcal{X}^n$ as follows. Start with any initial state. Whenever clock i rings, replace the ith coordinate of the process by an independently generated random number from the distribution μ. It is clear why this process may be called an 'independent flips' process. It is also clear that μ^n is the equilibrium measure for the process.

1 The Independent Flips Semigroup

Let us now compute the semigroup for this Markov process. Let $X = (X_1, \ldots, X_n)$ be a random vector in $\mathcal{X}^n$ with distribution μ^n. A moment's thought shows that

$$P_t f(x) = e^{-nt} f(x) + \sum_{k=1}^{n} (1-e^{-t})^k e^{-(n-k)t} \sum_{1 \le i_1 < i_2 < \cdots < i_k \le n} \mathbb{E}\big(f(Y_{i_1,\ldots,i_k})\big),$$

where $Y_{i_1,\ldots,i_k}$ is the random vector whose i_jth coordinate is X_{i_j}, $j = 1, \ldots, k$, and ith coordinate is x_i for all $i \notin \{i_1, \ldots, i_k\}$.

From this one easily computes the generator L of the process:

$$Lf(x) = \partial_t P_t f(x)|_{t=0} = \sum_{i=1}^{n} \big(\mathbb{E}(f(Y_i)) - f(x)\big),$$

where, as above, $Y_i = (x_1, \ldots, x_{i-1}, X_i, x_{i+1}, \ldots, x_n)$.

S. Chatterjee, *Superconcentration and Related Topics*,
Springer Monographs in Mathematics, DOI 10.1007/978-3-319-03886-5_7,
© Springer International Publishing Switzerland 2014

It is equally easy to compute the Dirichlet form. Let X' be an independent copy of the vector X. For each i, let

$$X^{\{i\}} = (X_1, \ldots, X_{i-1}, X'_i, X_{i+1}, \ldots, X_n).$$

The formula for L shows that the Dirichlet form is given by

$$\mathcal{E}(f,g) = -(f, Lg) = \sum_{i=1}^n \mathbb{E}\big[f(X)\big(g(X) - g(X^{\{i\}})\big)\big].$$

The Dirichlet form can be expressed a bit differently as follows. Take any i. Note that the pair $(X, X^{\{i\}})$ has the same distribution as the pair $(X^{\{i\}}, X)$. Thus,

$$\mathbb{E}\big[f(X)\big(g(X) - g(X^{\{i\}})\big)\big] = \mathbb{E}\big[f(X^{\{i\}})\big(g(X^{\{i\}}) - g(X)\big)\big].$$

Averaging the two sides, we get

$$\mathbb{E}\big[f(X)\big(f(X) - f(X^{\{i\}})\big)\big] = \frac{1}{2}\mathbb{E}\big[\big(f(X) - f(X^{\{i\}})\big)\big(g(X) - g(X^{\{i\}})\big)\big].$$

Therefore,

$$\mathcal{E}(f,g) = \frac{1}{2}\sum_{i=1}^n \mathbb{E}\big[\big(f(X) - f(X^{\{i\}})\big)\big(g(X) - g(X^{\{i\}})\big)\big].$$

The following theorem shows that any independent flips semigroup satisfies a Poincaré inequality with optimal constant 1. This is commonly known as the Efron-Stein inequality, proved by Efron and Stein (1981) and generalized by Steele (1986).

Theorem 7.1 (Efron-Stein inequality) *Take any product measure μ^n on a product space $\mathcal{X}^n$ and consider the independent flips semigroup P_t associated with the process. Then P_t satisfies a Poincaré inequality with optimal constant 1. In other words, for any $f \in L^2(\mu^n)$, $\mathrm{Var}_{\mu^n}(f) \leq \mathcal{E}(f,f)$.*

Proof Let X, X' and $X^{\{i\}}$ be as above. Let

$$X^{(i)} := (X'_1, \ldots, X'_i, X_{i+1}, \ldots, X_n).$$

Then

$$\mathrm{Var}_{\mu^n}(f) = \mathbb{E}\big[f(X)\big(f(X) - f(X')\big)\big]$$

$$= \sum_{i=1}^n \mathbb{E}\big[f(X)\big(f(X^{(i-1)}) - f(X^{(i)})\big)\big].$$

Take any i. In the expression $f(X)(f(X^{(i-1)}) - f(X^{(i)})))$ if we switch the random variables X_i and X'_i, the expectation should remain the same. That is,

$$\mathbb{E}[f(X)(f(X^{(i-1)}) - f(X^{(i)}))] = \mathbb{E}[f(X^{\{i\}})(f(X^{(i)}) - f(X^{(i-1)}))].$$

Averaging the two sides gives

$$\mathbb{E}[f(X)(f(X^{(i-1)}) - f(X^{(i)}))]$$
$$= \frac{1}{2}\mathbb{E}[(f(X) - f(X^{\{i\}}))(f(X^{(i-1)}) - f(X^{(i)}))].$$

Applying the Cauchy-Schwarz inequality gives

$$\mathbb{E}[(f(X) - f(X^{\{i\}}))(f(X^{(i-1)}) - f(X^{(i)}))]$$
$$\leq \left(\mathbb{E}[(f(X) - f(X^{\{i\}}))^2]\right)^{1/2} \left(\mathbb{E}[(f(X^{(i-1)}) - f(X^{(i)}))^2]\right)^{1/2}.$$

To complete the proof, note that the pair $(X, X^{\{i\}})$ has the same distribution as the pair $(X^{(i-1)}, X^{(i)})$. □

The Efron-Stein inequality was generalized to a very useful exponential inequality by Boucheron (2003).

2 Hypercontractivity for Independent Flips

Not all independent flips semigroups are hypercontractive. However, in the special case where the space $\mathcal{X}$ has exactly two elements and μ puts equal mass on the two points, the independent flips semigroup is hypercontractive. This is the famous Bonami-Beckner inequality, proved independently by Bonami (1970) and Beckner (1975).

Theorem 7.2 (Bonami-Beckner inequality) *Let μ be the uniform distribution on a two-point space $\mathcal{X}$, and let P_t be the independent flips semigroup with equilibrium measure μ^n on $\mathcal{X}^n$. Then for any $1 < p < \infty$, $t \geq 0$, and $f \in L^p(\mu^n)$,*

$$\|P_t f\|_{L^{q(t)}(\mu^n)} \leq \|f\|_{L^p(\mu^n)},$$

where $q(t) = 1 + (p-1)e^{2t}$.

Continuing with the case of the two-point space $\mathcal{X} = \{0, 1\}$ with uniform measure μ, define the 'discrete derivative' in the ith coordinate as

$$\Delta_i f(x) := f(x_1, \ldots, x_n) - f(x_1, \ldots, x_{i-1}, 1 - x_i, x_{i+1}, \ldots, x_n).$$

Talagrand's original L^1-L^2 inequality was formulated on this space.

Theorem 7.3 (Talagrand 1994) *For any $f \in L^2(\mu^n)$,*

$$\operatorname{Var}_{\mu^n}(f) \leq C \sum_{i=1}^{n} \frac{\|\Delta_i f\|_{L^2}^2}{1 + \log(\|\Delta_i f\|_{L^2}/\|\Delta_i f\|_{L^1})}$$

where C is a universal constant.

The following are left as exercises for the reader.

Exercise 7.4 Derive Talagrand's inequality from the Bonami-Beckner inequality.

Exercise 7.5 Prove superconcentration in first-passage percolation if each edge weight is either a or b with equal probability, where $0 < a < b$.

3 Chaos Under Independent Flips

The goal of this section is to prove the following theorem about the chaotic nature of the Sherrington-Kirkpatrick model under independent flips. Recall the notation and terminology of Chap. 1, Sect. 2.2.

Theorem 7.6 (Theorem 1.3 in Chatterjee 2009) *Consider the SK model at inverse temperature β. Take any integer n and $p \in [0, 1]$. Suppose that a randomly chosen fraction p of the couplings $(g_{ij})_{1 \leq i < j \leq n}$ are replaced by independent copies to give a perturbed Gibbs measure. Let σ^1 be chosen from the original Gibbs measure and σ^2 is chosen from the perturbed measure. Let $R_{1,2}$ be the overlap between the two configurations. Then*

$$\mathbb{E}(R_{1,2}^2) \leq \frac{C(\beta)}{p \log n},$$

where $C(\beta)$ depends only on β.

Theorem 7.6 is a corollary of the following general theorem about chaos under independent flips. This result and its proof are inspired by Lemma 2.3 in Chatterjee (2008a); we follow the same notation as in Chatterjee (2008a). Let $X = (X_1, \ldots, X_n)$ be a vector of independent random variables with $\operatorname{Var}(X_i) = 1$ for each i. Let X' be an independent copy of X. For any $A \subseteq [n] := \{1, \ldots, n\}$, let X^A be the vector whose ith component is

$$X_i^A := \begin{cases} X_i' & \text{if } i \in A, \\ X_i & \text{if } i \notin A. \end{cases}$$

Let $f : \mathbb{R}^n \to \mathbb{R}$ be a twice differentiable function. Let $\partial_i f$ and $\partial_i^2 f$ be the first and second partial derivatives of f in the ith coordinate.

3 Chaos Under Independent Flips

Theorem 7.7 *Suppose ϵ and δ are constants such that for all i, $|\partial_i f| \leq \delta$ and $|\partial_i^2 f| \leq \epsilon$ everywhere in the closed convex hull of the support of X. Fix $0 \leq k \leq n$, and let A be a subset of $[n]$, chosen uniformly at random from the collection of all subsets of size k. Define X^A as above. Let $\gamma := \max_i \mathbb{E}|X_i - X_i'|^3$. Then*

$$\mathbb{E}\left(\sum_{i=1}^n \partial_i f(X) \partial_i f(X^A)\right) \leq \frac{n+1}{k+1}\mathrm{Var}(f(X)) + \frac{3n\delta\epsilon\gamma}{2}. \qquad (7.1)$$

The proof of Theorem 7.7 is divided into a series of lemmas. First, to simplify notation, we will write f^A for $f(X^A)$. When $A = \emptyset$, we will simply write f. For any i and A such that $i \notin A$, let

$$\Delta_i f^A := f^A - f^{A \cup \{i\}}.$$

As usual, when $A = \emptyset$, we will simply write $\Delta_i f$. Let $\mathcal{A}_{k,i}$ denote the collection of all subsets of $[n]\setminus\{i\}$ of size k. For $0 \leq k \leq n-1$, define

$$T_k := \sum_{i=1}^n \frac{1}{\binom{n-1}{k}} \sum_{A \in \mathcal{A}_{k,i}} \mathbb{E}(\Delta_i f \Delta_i f^A).$$

The above quantity is a discrete proxy for the left-hand side in (7.1). Our first result is an exact formula for the variance in terms of $T_0, \ldots, T_{n-1}$. This is actually a restatement of Lemma 2.3 from Chatterjee (2008a).

Lemma 7.8 *We have*

$$\mathrm{Var}(f) = \frac{1}{2n}\sum_{k=0}^{n-1} T_k.$$

Proof By exchangeability of X_i and X_i', it is easy to see that the pair $(f, \Delta_i f^A)$ has the same distribution as the pair $(f^{\{i\}}, -\Delta_i f^A)$, and therefore

$$\mathbb{E}(\Delta_i f \Delta_i f^A) = \mathbb{E}(f \Delta_i f^A) - \mathbb{E}(f^{\{i\}} \Delta_i f^A) = 2\mathbb{E}(f \Delta_i f^A). \qquad (7.2)$$

We claim that

$$\frac{1}{n}\sum_{i=1}^n \sum_{k=0}^{n-1} \frac{1}{\binom{n-1}{k}} \sum_{A \in \mathcal{A}_{k,i}} \Delta_i f^A = f - f^{[n]}. \qquad (7.3)$$

To see this, consider any $B \subseteq [n]$ such that $B \neq \emptyset$ and $B \neq [n]$. Let $k = |B|$. On the left-hand side in the above display, if we write out the definition of $\Delta_i f^A$ as $f^A - f^{A \cup \{i\}}$ and regroup terms, then the coefficient of f^B in the expansion is

$$\frac{1}{n\binom{n-1}{k}}(n-k) - \frac{1}{n\binom{n-1}{k-1}}k = 0.$$

Similarly, the coefficient of f is 1 and the coefficient of $f^{[n]}$ is -1. This proves (7.3). Combining (7.3) with (7.2), we see that

$$\mathrm{Var}(f) = \mathbb{E}\big(f(f - f^{[n]})\big) = \frac{1}{2n} \sum_{i=1}^{n} \sum_{k=0}^{n-1} \frac{1}{\binom{n-1}{k}} \sum_{A \in \mathcal{A}_{k,i}} \mathbb{E}(\Delta_i f \Delta_i f^A).$$

This completes the proof of the lemma. □

Our next lemma is a monotonicity property of the T_k's.

Lemma 7.9 *We have* $T_0 \geq T_1 \geq \cdots \geq T_{n-1} \geq 0$.

Proof Take any A and $i \notin A$. It is easy to see that given $(X_j)_{j \notin A}$ and X'_i, the random variables $\Delta_i f$ and $\Delta_i f^A$ are i.i.d. Therefore,

$$\mathbb{E}(\Delta_i f \Delta_i f^A) = \mathbb{E}\big(\big(\mathbb{E}(\Delta_i f \mid (X_j)_{j \notin A}, X'_i)\big)^2\big).$$

From this and Jensen's inequality, it is clear that $\mathbb{E}(\Delta_i f \Delta_i f^A) \geq 0$, and for any $A \subseteq B \subseteq [n] \setminus \{i\}$,

$$\mathbb{E}(\Delta_i f \Delta_i f^A) \geq \mathbb{E}(\Delta_i f \Delta_i f^B).$$

Thus, if $k := |A| \leq n - 2$, we have

$$\mathbb{E}(\Delta_i f \Delta_i f^A) \geq \frac{1}{n-k-1} \sum \mathbb{E}(\Delta_i f \Delta_i f^B),$$

where the sum is taken over all B such that $B = A \cup \{j\}$ for some $j \notin A \cup \{i\}$. Since any $B \in \mathcal{A}_{k+1,i}$ can be obtained by adding one element to A for exactly $k+1$ many $A \in \mathcal{A}_{k,i}$, we have

$$\sum_{A \in \mathcal{A}_{k,i}} \mathbb{E}(\Delta_i f \Delta_i f^A) \geq \frac{k+1}{n-k-1} \sum_{B \in \mathcal{A}_{k+1,i}} \mathbb{E}(\Delta_i f \Delta_i f^B).$$

This can be rewritten as

$$\frac{1}{\binom{n-1}{k}} \sum_{A \in \mathcal{A}_{k,i}} \mathbb{E}(\Delta_i f \Delta_i f^A) \geq \frac{1}{\binom{n-1}{k+1}} \sum_{B \in \mathcal{A}_{k+1,i}} \mathbb{E}(\Delta_i f \Delta_i f^B).$$

This completes the proof of the lemma. □

Combining Lemmas 7.8 and 7.9, we easily get the following discrete version of Theorem 7.7.

Lemma 7.10 *For each* $0 \leq k \leq n - 1$,

$$T_k \leq \frac{2n \, \mathrm{Var}(f)}{k+1}.$$

3 Chaos Under Independent Flips

Proof Since $T_0 \geq T_1 \geq \cdots \geq T_{n-1} \geq 0$, and

$$\text{Var}(f) = \frac{1}{2n} \sum_{k=0}^{n-1} T_k,$$

it follows that for each $0 \leq k \leq n-1$,

$$T_k \leq \frac{1}{k+1} \sum_{r=0}^{k} T_r \leq \frac{2n \, \text{Var}(f)}{k+1}.$$

This completes the proof of the lemma. $\square$

Finally, we are ready to prove Theorem 7.7. This involves replacing the discrete derivatives in Lemma 7.10 with continuous derivatives, and incurring a small error along the way.

Proof of Theorem 7.7 Since $|\partial_i f| \leq \delta$ and $|\partial_i^2 f| \leq \epsilon$ everywhere on the closed convex hull of the support of X, by Taylor expansion we have

$$|\Delta_i f^A| \leq |X_i - X_i'|\delta, \qquad |\Delta_i f^A - (X_i - X_i')\partial_i f^A| \leq \frac{\epsilon}{2}(X_i - X_i')^2.$$

Thus,

$$\begin{aligned}
|\mathbb{E}(\Delta_i f \Delta_i f^A) &- \mathbb{E}((X_i - X_i')^2 \partial_i f \partial_i f^A)| \\
&\leq |\mathbb{E}((\Delta_i f - (X_i - X_i')\partial_i f)\Delta_i f^A)| \\
&\quad + |\mathbb{E}((X_i - X_i')\partial_i f(\Delta_i f^A - (X_i - X_i')\partial_i f^A))| \\
&\leq \delta\epsilon \mathbb{E}|X_i - X_i'|^3.
\end{aligned} \qquad (7.4)$$

Now let X_i'' be another independent copy of X_i, that is also independent of X_i'. Let $\widetilde{\partial_i f}$ denote $\partial_i f$ with X_i replaced by X_i'' and define $\widetilde{\partial_i f}^A$ similarly. Since $\text{Var}(X_i) = 1$ and $(X_i - X_i')^2$ is independent of $\widetilde{\partial_i f}\widetilde{\partial_i f}^A$, we have

$$\mathbb{E}((X_i - X_i')^2 \widetilde{\partial_i f}\widetilde{\partial_i f}^A) = 2\mathbb{E}(\widetilde{\partial_i f}\widetilde{\partial_i f}^A) = 2\mathbb{E}(\partial_i f \partial_i f^A).$$

Again,

$$|\partial_i f \partial_i f^A - \widetilde{\partial_i f}\widetilde{\partial_i f}^A| \leq 2\delta\epsilon |X_i - X_i''|.$$

Combining the last two observations, we get

$$\begin{aligned}
|\mathbb{E}((X_i - X_i')^2 \partial_i f \partial_i f^A) - 2\mathbb{E}(\partial_i f \partial_i f^A)| &\leq 2\delta\epsilon \mathbb{E}((X_i - X_i')^2 |X_i - X_i''|) \\
&\leq 2\delta\epsilon \mathbb{E}|X_i - X_i'|^3.
\end{aligned}$$

And now, combining the above bound with (7.4), we have

$$2\,\mathbb{E}\big(\partial_i f \partial_i f^A\big) \leq \mathbb{E}\big(\Delta_i f \Delta_i f^A\big) + 3\delta\epsilon\mathbb{E}\big|X_i - X'_i\big|^3. \tag{7.5}$$

We also have to consider the case when $i \in A$. Let $B = A\setminus\{i\}$. Then by Jensen's inequality we have

$$\mathbb{E}\big(\partial_i f \partial_i f^A\big) = \mathbb{E}\big((\mathbb{E}(\partial_i f \mid (X_j)_{j \notin A}))^2\big)$$
$$\leq \mathbb{E}\big((\mathbb{E}(\partial_i f \mid (X_j)_{j \notin B}))^2\big) = \mathbb{E}\big(\partial_i f \partial_i f^B\big). \tag{7.6}$$

Now take $1 \leq k \leq n-1$ and let $\mathcal{A}_k$ denote the set of all subsets of $[n]$ of size k. Using (7.5) and (7.6), we get

$$\sum_{i=1}^{n} \sum_{A \in \mathcal{A}_k} \mathbb{E}\big(\partial_i f \partial_i f^A\big) = \sum_{i=1}^{n} \left(\sum_{A \in \mathcal{A}_{k,i}} \mathbb{E}\big(\partial_i f \partial_i f^A\big) + \sum_{A \in \mathcal{A}_{k-1,i}} \mathbb{E}\big(\partial_i f \partial_i f^{A \cup \{i\}}\big) \right)$$

$$\leq \sum_{i=1}^{n} \left(\sum_{A \in \mathcal{A}_{k,i}} \mathbb{E}\big(\partial_i f \partial_i f^A\big) + \sum_{A \in \mathcal{A}_{k-1,i}} \mathbb{E}\big(\partial_i f \partial_i f^A\big) \right)$$

$$\leq \frac{1}{2}\binom{n-1}{k} T_k + \frac{1}{2}\binom{n-1}{k-1} T_{k-1} + \frac{n}{2}\binom{n}{k} 3\delta\epsilon\gamma.$$

From this and Lemma 7.10, we conclude that for $1 \leq k \leq n-1$,

$$\frac{1}{\binom{n}{k}} \sum_{i=1}^{n} \sum_{A \in \mathcal{A}_k} \mathbb{E}\big(\partial_i f \partial_i f^A\big) \leq \frac{n-k}{2n} T_k + \frac{k}{2n} T_{k-1} + \frac{3n\delta\epsilon\gamma}{2}$$

$$\leq \frac{n+1}{k+1} \mathrm{Var}(f) + \frac{3n\delta\epsilon\gamma}{2}.$$

The same conclusion can be drawn for $k = 0$ and $k = n$ by defining $T_{-1} = T_n = 0$ and verifying that all steps hold. This completes the proof. □

Proof of Theorem 7.6 Fix β, and define $f = n^{-1/2} F_n(\beta)$, where $F_n(\beta)$ is the free energy of the SK model defined in Chap. 1, Sect. 1.3. Consider f as a function of the disorder $g = (g_{ij})_{1 \leq i \leq j \leq n}$. Let g' be an independent copy of g, and define g^A as we defined X^A in Theorem 7.7. Let $k = pN$ (assume for simplicity that k is an integer), and define a perturbed Hamiltonian using the disorder g^A, where A is chosen uniformly at random from the set of all subsets of $\{(i, j)\}_{1 \leq i, j \leq n}$ of size k.

Let σ^1 be sampled from the original Gibbs measure, and σ^2 from the perturbed Gibbs measure. An easy verification shows that

$$\sum_{i,j} \partial_{ij} f(g) \partial_{ij} f\big(g^A\big) = \frac{1}{2}\langle R_{1,2}^2 \rangle - \frac{1}{2n},$$

where $\partial_{ij} f$ is the partial derivative of f in the (i,j)th coordinate. On the other hand, by Theorem 1.6 we know that

$$\mathrm{Var}(f(g)) \le \frac{C(\beta)}{\log n}.$$

Finally, note that for any (i,j),

$$\partial_{ij} f = \frac{\langle \sigma_i \sigma_j \rangle}{n}, \qquad \partial_{ij}^2 f = \frac{\beta(1-\langle \sigma_i \sigma_j \rangle^2)}{n^{3/2}}.$$

Therefore, we can take $\delta = n^{-1}$ and $\epsilon = \beta n^{-3/2}$ while applying Theorem 7.7. Using all the above information, we can now apply Theorem 7.7 to conclude that

$$\mathbb{E}\langle R_{1,2}^2 \rangle \le \frac{C(\beta)}{p \log n} + C(\beta) n^{-1/2},$$

where $C(\beta)$ depends only on β. Since $p \in (0,1)$, we can ignore the second term on the right by appropriately increasing the value of $C(\beta)$. □

Chapter 8
Extremal Fields

Let $g = (g_1, \ldots, g_n)$ be a centered Gaussian field with covariance matrix R, as in Sect. 1.4 of Chap. 1. Throughout this section, suppose that $R(i,i) \leq 1$ for each i. Our object of interest is

$$M := \max_{1 \leq i \leq n} g_i.$$

If $R(i,j) = 0$ for each $i \neq j$ and $R(i,i) = 1$ for each i, then it is well-known that $M \approx \sqrt{2 \log n}$ with high probability (see Appendix A). Moreover, the expected value of M is always bounded above by $\sqrt{2 \log n}$, irrespective of the correlation structure (again, see Appendix A). In the presence of correlations, the expected value of the maximum usually decreases to a smaller multiple of $\sqrt{2 \log n}$.

It is a surprising fact that a number of well-known Gaussian fields satisfy $\mathbb{E}(M) \approx \sqrt{2 \log n}$, in spite of a lot of dependence among the coordinates. In Chatterjee (2008b), the term 'extremal' was used to describe such fields. Examples include branching random walks and the discrete Gaussian free field.

1 Superconcentration in Extremal Fields

The following theorem shows that the maximum of an extremal Gaussian field must necessarily be superconcentrated.

Theorem 8.1 (Theorem 5.1 in Chatterjee 2008b) *Let $g = (g_1, \ldots, g_n)$ be a centered Gaussian field satisfying* $\mathrm{Var}(g_i) \leq 1$ *for each i. Let*

$$\alpha := \frac{\mathbb{E}(\max_{1 \leq i \leq n} g_i)}{\sqrt{2 \log n}}.$$

Then there is a universal constant C such that

$$\mathrm{Var}\left(\max_{1 \leq i \leq n} g_i\right) \leq C\sqrt{1 - \alpha} + C\left(\frac{\log \log n}{\log n}\right)^{1/4}.$$

This result was used in Chatterjee (2008b) to prove superconcentration of the maximum of the discrete Gaussian free field, a result that has been vastly refined in recent years (see e.g. Bramson and Zeitouni 2009, Ding 2011, Ding and Zeitouni 2012). It was also used to prove the superconcentration of the ground state energy in a certain class of mean-field spin glass models in Chatterjee (2008b). We will not go into the details of those in this monograph.

The proof of Theorem 8.1 proceeds roughly along the following steps:

(1) Let $m := \mathbb{E}(\max_{1\leq i \leq n} g_i)$. Let g^t and I^t be defined as in Sect. 2.3 of Chap. 1. We claim that for any $t \geq 0$, $g_{I^t} \approx e^{-t}m$ with high probability.
(2) Next, fix t and let $D = \{i : g_i \approx e^{-t}m\}$, where the meaning of $\approx$ will be made precise in the proof.
(3) A simple first moment bound shows that with high probability, $|D| \lesssim n^{1-e^{-2t}}$.
(4) Next fix some $r \geq 0$ and let $B = \{i : R(I_0, i) \geq r\}$. The key observation is that since g' and g are independent and $\mathrm{Var}(g_i) \leq 1$ for all i, $\mathrm{Var}(g'_i - rg'_{I_0} \mid g) \leq 1 - r^2$ for each $i \in B$.
(5) From the previous step, it follows that

$$\mathbb{E}\Big(\max_{i \in B \cap D} g'_i \mid g\Big) = \mathbb{E}\Big(\max_{i \in B \cap D} (g'_i - rg'_{I_0}) \mid g\Big)$$

$$\lesssim \sqrt{(1-r^2) 2 \log |D|}.$$

(6) Combining this with the bound $|D| \lesssim n^{1-e^{-2t}}$ gives

$$\max_{i \in B \cap D} g_i^t \leq e^{-t} \max_{i \in B \cap D} g_i + \sqrt{1 - e^{-2t}} \max_{i \in B \cap D} g'_i$$

$$\lesssim e^{-2t}\alpha\sqrt{2 \log n} + (1 - e^{-2t})\sqrt{2(1-r^2) \log n}.$$

(7) Thus, if $\sqrt{1-r^2} < \alpha$, or in other words $r > \sqrt{1-\alpha^2}$, we cannot have that $\max_{i \in B \cap D} g_i^t \approx \alpha\sqrt{2 \log n}$. If this does not happen, then $I^t \notin B \cap D$. But we have already stated that $I^t \in D$ with high probability. Therefore, whenever $r > \sqrt{1-\alpha^2}$, we have that $I^t \in D \backslash B$ with high probability, which implies that $R(I^0, I^t) < r$ with high probability.
(8) Roughly, this justifies the $\sqrt{1-\alpha}$ term in the statement of the theorem. The second term arises from our attempts at making the above sketch rigorous, which is a somewhat technically involved task.

Let us now begin the actual proof. The first step, as mentioned above, is to show that $g_{I^t} \approx e^{-t}m$. This is made precise in the following lemma.

Lemma 8.2 *Take any $t \geq 0$. Then for any $x \geq 0$, we have*

$$\mathbb{P}\big(|g_{I^t} - e^{-t}m| \geq x\big) \leq 4e^{-x^2/4}.$$

1 Superconcentration in Extremal Fields

Proof For notational convenience, let $a = e^{-t}$, $b = \sqrt{1-e^{-2t}}$, $h := g^t$, and $h' := bg - ag'$. Then h and h' are independent and have the same distribution as g (verified by computing correlations), and

$$g = ah + bh'. \tag{8.1}$$

Since $a + b \leq \sqrt{2}$, by inequality (A.7) of Appendix A and the independence of h and h', we have

$$\mathbb{P}(|g_{I^t} - am| \geq x) \leq \mathbb{P}(a|h_{I^t} - m| \geq ax/\sqrt{2}) + \mathbb{P}(b|h'_{I^t}| \geq bx/\sqrt{2})$$
$$\leq 2e^{-x^2/4} + 2e^{-x^2/4}.$$

This completes the proof. □

Proof of Theorem 8.1 As in Lemma 8.2, let $a = e^{-t}$, $b = \sqrt{1 - e^{-2t}}$, and $h = g^t$. Note that since $|R(i, j)| \leq 1$ for all i, j and C can be chosen as large as we like, it suffices to prove the theorem assuming that n is larger than some fixed threshold and that

$$\frac{1}{b}\left(\frac{\log\log n}{\alpha^2 \log n}\right)^{1/4} \leq \frac{1}{100}. \tag{8.2}$$

For the same reason, there is no loss of generality in assuming that $m \geq 2$. With that assumption, define

$$\delta := \frac{\sqrt{\log(m^4/2)}}{m} \in (0, 1), \tag{8.3}$$

and let

$$D := \{i : |g_i - am| \leq \delta m\}.$$

By Lemma 8.2,

$$\mathbb{P}(I^t \notin D) \leq 4e^{-\delta^2 m^2/4}. \tag{8.4}$$

Now, if $a \geq \delta$, then by the Gaussian tail bound from Appendix A,

$$\mathbb{E}|D| \leq \sum_i \mathbb{P}(g_i \geq (a - \delta)m)$$
$$\leq ne^{-(a-\delta)^2 m^2/2} = n^{1-(a-\delta)^2 \alpha^2}.$$

Therefore, if we define

$$\zeta := \begin{cases} 1 - (a - \delta)^2 \alpha^2 + \delta^2 & \text{if } a \geq \delta, \\ 1 & \text{if } a < \delta, \end{cases}$$

then in all cases we have

$$\mathbb{P}(|D| > n^\zeta) \leq n^{-\delta^2}. \tag{8.5}$$

Define
$$\gamma := \frac{\alpha(b^2 - 2\delta)}{b\sqrt{\zeta}}.$$

It is easy to verify using (8.2) that $b^2 \geq 2\delta$ and hence $\gamma \geq 0$. Again,
$$\frac{\alpha(b^2 - 2\delta)}{b\sqrt{\zeta}} \leq \frac{b^2 - 2\delta}{b\sqrt{1-a^2}} = \frac{b^2 - 2\delta}{b^2} \leq 1.$$

Thus, $\gamma \in [0, 1]$. Let $r := \sqrt{1 - \gamma^2}$, and define the random set
$$B := \{i : R(I^0, i) \geq r\}.$$

Note that
$$\mathbb{P}(R(I^0, I^t) \geq r) = \mathbb{P}(I^t \in B) \leq \mathbb{P}(I^t \in B \cap D) + \mathbb{P}(I^t \notin D). \quad (8.6)$$

Let $\mathbb{E}^0$ and Var^0 denote the conditional expectation and variance given g. Since $R(i, i) \leq 1$ and $R(I^0, i) \geq r$ for all $i \in B$ and g' is independent of g, we have
$$\mathrm{Var}^0(g'_i - rg'_{I^0}) \leq 1 + r^2 - 2r^2 = \gamma^2.$$

Again, $\mathbb{E}^0(g'_{I^0}) = 0$. Thus, if $B \cap D \neq \emptyset$, then by Lemma A.1 from Appendix A we have
$$\mathbb{E}^0\left(\max_{i \in B \cap D} g'_i\right) = \mathbb{E}^0\left(\max_{i \in B \cap D} (g'_i - rg'_{I^0})\right)$$
$$\leq \gamma\sqrt{2\log|B \cap D|} \leq \gamma\sqrt{2\log|D|}.$$

Combined with inequality (A.7), this implies that if $B \cap D \neq \emptyset$, then for all $x \geq 0$,
$$\mathbb{P}^0\left(\max_{i \in B \cap D} g'_i \geq \gamma\sqrt{2\log|D|} + x\right) \leq e^{-x^2/2}. \quad (8.7)$$

Clearly, the inequality holds even if we relax the condition $B \cap D \neq \emptyset$ to just $D \neq \emptyset$, interpreting the maximum of an empty set as $-\infty$. Since $g_i \leq (a + \delta)m$ for $i \in D$ and $h = ag + bg'$, we have
$$\max_{i \in B \cap D} h_i \leq a \max_{i \in B \cap D} g_i + b \max_{i \in B \cap D} g'_i$$
$$\leq a(a + \delta)m + b \max_{i \in B \cap D} g'_i.$$

Thus, from (8.7), we see that whenever $D \neq \emptyset$,
$$\mathbb{P}^0\left(\max_{i \in B \cap D} h_i \geq a(a + \delta)m + b\gamma\sqrt{2\log|D|} + bx\right) \leq e^{-x^2/2}. \quad (8.8)$$

1 Superconcentration in Extremal Fields

Putting

$$m' := a(a+\delta)m + b\gamma\sqrt{2\zeta \log n}$$
$$= \left(a(a+\delta) + \frac{b\gamma\sqrt{\zeta}}{\alpha}\right)m$$
$$= (a^2 + a\delta + b^2 - 2\delta)m = (1 + a\delta - 2\delta)m,$$

and using (8.8) and (8.5) we get

$$\mathbb{P}\left(\max_{i \in B \cap D} h_i \geq m' + bx\right) \leq \mathbb{E}\left(\mathbb{P}^0\left(\max_{i \in B \cap D} h_i \geq m' + bx\right); 1 \leq |D| \leq n^{\zeta}\right)$$
$$+ \mathbb{E}\left(\mathbb{P}^0\left(\max_{i \in B \cap D} h_i \geq m' + bx\right); |D| > n^{\zeta}\right)$$
$$\leq e^{-x^2/2} + n^{-\delta^2}. \tag{8.9}$$

Now

$$m - m' = (2-a)\delta m \geq \delta m. \tag{8.10}$$

In particular, $m' \leq m$. Let $x := (m-m')/2b$. Then by (8.9) and inequality (A.7) we have

$$\mathbb{P}(I^t \in B \cap D) \leq \mathbb{P}\left(\max_{i \in B \cap D} h_i \geq m' + bx\right) + \mathbb{P}\left(\max_{i \in S} h_i \leq m - bx\right)$$
$$\leq e^{-x^2/2} + n^{-\delta^2} + e^{-b^2 x^2/2}$$
$$\leq 2e^{-(m-m')^2/8} + n^{-\delta^2}. \tag{8.11}$$

Thus, by (8.4), (8.6), (8.10), and (8.11), we have

$$\mathbb{P}(R(I^0, I^t) \geq r) \leq 6e^{-\delta^2 m^2/8} + n^{-\delta^2} \leq 7e^{-\delta^2 m^2/8}. \tag{8.12}$$

Now, if $a \geq \delta$, then

$$r^2 = \frac{b^2\zeta - \alpha^2(b^4 - 4b^2\delta + 4\delta^2)}{b^2\zeta}$$
$$= \frac{b^2(1 - a^2\alpha^2 + 2a\delta\alpha^2 - \delta^2\alpha^2 + \delta^2) - \alpha^2(b^4 - 4b^2\delta + 4\delta^2)}{b^2\zeta}$$
$$\leq \frac{(1-\alpha^2)b^2 + 2b^2 a\delta\alpha^2 + b^2\delta^2 + 4\alpha^2 b^2\delta}{b^4}$$
$$\leq \frac{1 - \alpha^2 + C\delta}{b^2},$$

where C denotes a universal constant. Again, if $a < \delta$,

$$r^2 = \frac{b^2 - \alpha^2(b^4 - 4b^2\delta + 4\delta^2)}{b^2}$$
$$\leq 1 - \alpha^2 b^2 + 4\alpha^2 \delta$$
$$= 1 - \alpha^2 + \alpha^2 a^2 + 4\alpha^2 \delta \leq 1 - \alpha^2 + 5\delta.$$

Therefore in all cases we have

$$\mathbb{E}(R(I^0, I^t)) \leq r + \mathbb{P}(R(I^0, I^t) \geq r)$$
$$\leq \left(\frac{1 - \alpha^2 + C\delta}{b^2}\right)^{1/2} + Ce^{-\delta^2 m^2/8}.$$

From the definition (8.3) of δ we have

$$e^{-\delta^2 m^2/8} = \frac{2^{1/8}}{m^{1/2}}.$$

Since $\sqrt{x+y} \leq \sqrt{x} + \sqrt{y}$ and $\alpha \leq 1$, we get

$$\mathbb{E}(R(I^0, I^t)) \leq \frac{\sqrt{2(1-\alpha)}}{b} + \frac{C(\log m)^{1/4}}{bm^{1/2}}.$$

Now, $m^{1/2} = \alpha^{1/2}(2\log n)^{1/4}$. Since $R(i,j) \leq 1$ for all i, j, we can put a large enough constant in front of the first term to remove the $\alpha^{1/2}$ from the denominator of the second term. Finally, the bound on $\mathrm{Var}(\max_{1 \leq i \leq n} g_i)$ comes from the identity (1.4) of Chap. 1 and the above bound on $\mathbb{E}(R(I^0, I^t))$. This completes the proof of the theorem. □

2 A Sufficient Condition for Extremality

The following theorem shows that when 'correlations are sufficiently small', one can sometimes establish extremality using some simple estimates.

Theorem 8.3 *Let $g = (g_i)_{1 \leq i \leq n}$ be a centered Gaussian field. Let $R(i,j) := \mathrm{Cov}(g_i, g_j)$. Suppose that $R(i,i) = 1$ for all i. Then there are positive universal constants C_1 and C_2 such that*

$$\mathbb{E}\left(\max_{1 \leq i \leq n} g_i\right) \geq \sqrt{2\log n} - C_1\left(\log\log n + \log \sum_{i,j \in S} n^{-2/(1+R(i,j))}\right)^{1/2}$$

and consequently by Theorem 8.1,

$$\mathrm{Var}\left(\max_{1 \leq i \leq n} g_i\right) \leq C_2\left(\frac{\log\log n + \log \sum_{i,j} n^{-2/(1+R(i,j))}}{\log n}\right)^{1/4}.$$

Proof In this proof, C will be any positive universal constant. The value of C may change from line to line. Define $M := \max_{1 \le i \le n} g_i$ and

$$N := |\{i : g_i \ge \sqrt{2\log n}\}|.$$

Then by the Mills ratio lower bound (A.1) from Appendix A and the assumption that $R(i,i) = 1$ for all i,

$$\mathbb{E}(N) \ge Cn \frac{e^{-\log n}}{\sqrt{2\log n}} = \frac{C}{\sqrt{2\log n}}.$$

Again,

$$\mathbb{E}(N^2) = \sum_{i,j} \mathbb{P}(g_i \ge \sqrt{2\log n}, \, g_j \ge \sqrt{2\log n})$$

$$\le \sum_{i,j} \mathbb{P}(g_i + g_j \ge 2\sqrt{2\log n})$$

$$\le \sum_{i,j} \exp\left(-\frac{4\log n}{\mathrm{Var}(g_i + g_j)}\right)$$

$$= \sum_{i,j} \exp\left(-\frac{2\log n}{1 + R(i,j)}\right).$$

Applying the Paley-Zygmund second moment method together with the above inequalities gives

$$\mathbb{P}(M \ge \sqrt{2\log n}) = \mathbb{P}(N > 0) \ge \frac{(\mathbb{E}(N))^2}{\mathbb{E}(N^2)} \ge \frac{C}{\sum_{i,j} n^{-2/(1+R(i,j))} \log n}.$$

Again, by Proposition A.7 we know that for any $x \ge 0$,

$$\mathbb{P}(M - \mathbb{E}(M) \ge x) \le e^{-x^2/2}.$$

Combining the last two inequalities gives the inequality for the expectation in the statement of the theorem. For the variance inequality, simply combine the expectation inequality with Theorem 8.1. □

3 Application to Spin Glasses

The Sherrington-Kirkpatrick model admits a popular generalization to the so-called p-spin models. The Hamiltonian for the p-spin model is defined as

$$H_{n,p}(\sigma) = -\frac{1}{n^{(p-1)/2}} \sum_{1 \le i_1, i_2, \ldots, i_p \le n} g_{i_1 i_2 \cdots i_p} \sigma_{i_1} \sigma_{i_2} \cdots \sigma_{i_p}, \quad (8.13)$$

where $(g_{i_1 i_2 \cdots i_p})_{1 \le i_1,\ldots,i_p \le n}$ is a fixed realization of i.i.d. standard Gaussian random variables. (Usually, the sum is taken over distinct $i_1,\ldots,i_p$. We take it over all $i_1,\ldots,i_p$ to avoid technical inconveniences.) The p-spin model was suggested by Derrida, and subsequently studied by Gross and Mézard (1984) and Gardner (1985).

A generalized version of the SK model that covers all p-spin models was considered by Talagrand in his celebrated paper on the Parisi formula Talagrand (2006). It is a linear combination of the p-spin energies over all p. Given a sequence of non-negative real numbers $c = (c_2, c_3, \ldots)$ such that

$$\sum_{p=2}^{\infty} c_p = 1, \tag{8.14}$$

define the Hamiltonian

$$H_{n,c}(\sigma) := \sum_{p=2}^{\infty} c_p^{1/2} H_{n,p}(\sigma), \tag{8.15}$$

where $H_{n,p}$ is the p-spin energy defined above in (8.13). The usual SK model corresponds to the sequence $(1, 0, 0, \ldots)$, and the p-spin model corresponds to the sequence that has 1 at the pth position and 0 elsewhere.

The ground state energy of the system is the minimum value of $H_{n,c}(\sigma)$ as σ ranges over $\{-1, 1\}^n$. The energy of the ground state is said to have fluctuation exponent ρ if it has fluctuations of order $n^{\rho + o(1)}$ as $n \to \infty$. A simple application of the Gaussian Poincaré inequality (2.5) shows that $\rho \le 1/2$ for any c satisfying (8.14). The objective of this section is to show that there are nontrivial sequences c where the fluctuation exponent may be proved to be strictly less than $1/2$. This is one of the few instances where superconcentration can be proved with a better than logarithmic improvement in the variance bound.

Theorem 8.4 *Let $I(x) := \frac{1}{2}((1+x)\log(1+x) + (1-x)\log(1-x))$, and $(c_p^*)_{p \ge 2}$ be constants such that for $x \in (-1, 1)$,*

$$\frac{I(x)}{2\log 2 - I(x)} = \sum_{p=2}^{\infty} c_p^* x^p.$$

Then $c_p^* \ge 0$ for all p and $\sum c_p^* = 1$. Suppose that $c = (c_p)_{p \ge 2}$ is any non-negative sequence such that $\sum c_p = 1$, and for all $r \ge 2$,

$$\sum_{p=2}^{r} c_p \le \sum_{p=2}^{r} c_p^*. \tag{8.16}$$

Then the ground state energy of the Hamiltonian (8.15) *has fluctuation exponent $\le 3/8$.*

3 Application to Spin Glasses

The proof of Theorem 8.4 is based on an application of Theorem 8.1. The minorizing condition (8.16) suffices to guarantee extremality of the energy landscape seen as a Gaussian field. We will actually prove a more general version of Theorem 8.4, with precise quantitative bounds. Fix n, and consider a centered Gaussian field $h = (h_\sigma)_{\sigma \in \{-1,1\}^n}$ satisfying

$$\mathrm{Cov}(h_\sigma, h_{\sigma'}) = \xi\left(\frac{\sigma \cdot \sigma'}{n}\right) \quad \text{for all } \sigma \in \{-1, 1\}^n,$$

where $\sigma \cdot \sigma' = \sum_{i=1}^n \sigma_i \sigma'_i$ is the usual inner product, and $\xi : [-1, 1] \to [-1, 1]$ is a function with $\xi(1) = 1$. The following theorem is a quantitative version of Theorem 8.4.

Theorem 8.5 *Let $I(x)$ be as in Theorem 8.4. Suppose that*

$$|\xi(x)| \leq \xi(|x|) \quad \text{and} \quad \xi(x) \leq \frac{I(x)}{2\log 2 - I(x)} \quad \text{for all } x \in (-1, 1). \quad (8.17)$$

Then

$$\mathrm{Var}\left(\max_\sigma h_\sigma\right) \leq C\left(\frac{\log n}{n}\right)^{1/4},$$

where C is a universal constant.

Before proving Theorem 8.5, let us first show that it implies Theorem 8.4. First of all, if we define $h_\sigma = -n^{-1/2} H_{n,c}(\sigma)$, then

$$\mathrm{Cov}(h_\sigma, h_{\sigma'}) = \sum_{p=2}^\infty c_p \left(\frac{\sigma \cdot \sigma'}{n}\right)^p.$$

Thus, we are in the setting of Theorem 8.5 with $\xi(x) = \sum c_p x^p$. So if we can show that (8.17) follows from (8.16), then Theorem 8.5 would imply that $\mathrm{Var}(\max_\sigma h_\sigma) \leq n^{-1/4+o(1)}$ and hence that

$$\mathrm{Var}\left(\min_\sigma H_{n,c}(\sigma)\right) \leq n^{3/4+o(1)},$$

which proves the claim. The implication of (8.17) from (8.16) is proved as follows. First, it is easy to verify that the power series for $I(x)$ has non-negative coefficients, and therefore so does

$$\frac{I(x)}{2\log 2 - I(x)} = \sum_{k=1}^\infty \left(\frac{I(x)}{2\log 2}\right)^k.$$

For each r, let $C_r = \sum_{p=2}^r c_p$ and $C_r^* = \sum_{p=2}^r c_p^*$, with $C_1 = C_1^* = 0$. The assumption (8.16) says that $C_r \leq C_r^*$ for each r. Thus for any $x \in (0, 1)$,

$$\xi(x) = \sum_{p=2}^{\infty} c_p x^p = \sum_{p=2}^{\infty} (C_p - C_{p-1}) x^p$$

$$= \sum_{p=2}^{\infty} (x^p - x^{p+1}) C_p$$

$$\leq \sum_{p=2}^{\infty} (x^p - x^{p+1}) C_p^* = \sum_{p=2}^{\infty} c_p^* x^p = \frac{I(x)}{2 \log 2 - I(x)}.$$

Since $|\xi(x)| \leq \xi(|x|)$ and I is symmetric, the inequality holds for $x \in (-1, 0]$ as well. This completes the argument for Theorem 8.4.

Proof of Theorem 8.5 We use Theorem 8.3. Let 1 denote the configuration of all 1's. Then by symmetry, we have

$$\sum_{\sigma, \sigma' \in \{-1,1\}^n} 2^{-2n/(1+\xi(\frac{\sigma \cdot \sigma'}{n}))} = 2^n \sum_{\sigma \in \{-1,1\}^n} \exp\left(-\frac{2n \log 2}{1 + \xi(\frac{1 \cdot \sigma}{n})}\right).$$

By the binomial theorem, we know that the number of configurations that have $\sum_{i=1}^n \sigma_i = k$ is exactly

$$\binom{n}{\frac{n+k}{2}},$$

which is interpreted as zero if k and n have different parity. Now, we have that for any $p \in [0, 1]$,

$$\binom{n}{\frac{n+k}{2}} p^{(n+k)/2} (1 - p)^{(n-k)/2} \leq 1.$$

Taking $p = (n+k)/2n$, we get

$$\binom{n}{\frac{n+k}{2}} \leq e^{n(\log 2 - I(k/n))},$$

where $I(x)$ is defined in the statement of the theorem. Again, the hypothesis implies that

$$\frac{2 \log 2}{1 + \xi(x)} \geq 2 \log 2 - I(x) \quad \text{for all } x \in [-1, 1].$$

Thus,

$$\sum_{\sigma, \sigma' \in \{-1,1\}^n} 2^{-2n/(1+\xi(\frac{\sigma \cdot \sigma'}{n}))} \leq 2^n \sum_{k=-n}^n \exp\left(-\frac{2n \log 2}{1 + \xi(k/n)}\right) \binom{n}{\frac{n+k}{2}} \leq Cn.$$

The proof now follows from Theorem 8.3. □

4 Application to the Discrete Gaussian Free Field

In this section we will use extremality to show that the maximum of the discrete Gaussian free field (DGFF) on an $n \times n$ grid (defined below) is superconcentrated. The massless Gaussian free field is an important mathematical object, inspiring a substantial amount of rigorous literature. It is essentially a higher dimensional analog of Brownian motion, where the dimension of time (rather than space) is higher than one. Although initially introduced as a toy model for the Ising interface, the topic has grown in its own right and has found important intersections with subjects as diverse as quantum gravity and stochastic Loewner evolutions. The DGFF is a finite approximation to the massless free field, just as random walk is a finite approximation of Brownian motion. For further motivation, definitions, and a review of the rigorous literature, we refer to Giacomin (2000) and the excellent survey of Sheffield (2007).

Let $V_n := \{0,\ldots,n-1\}^2$, and ∂V_n be the inner boundary, that is, the points in V_n which have a nearest neighbor outside. Let $\text{int}(V_n) := V_n \setminus \partial V_n$. The two-dimensional discrete Gaussian free field on V_n with zero boundary condition is defined as a family $\Phi_n = \{\phi_x\}_{x \in V_n}$ of centered Gaussian random variables with covariances given by the discrete Green's function of the (discrete) Laplacian on $\text{int}(V_n)$. This means, explicitly, that $\phi_x \equiv 0$ for $x \in \partial V_n$, and

$$\text{Cov}(\phi_x, \phi_y) = G_n(x,y) = \mathbb{E}_x\left(\sum_{i=0}^{\tau_{\partial V_n}} \mathbb{I}_{\{\eta_i = y\}}\right), \quad x, y \in \text{int}(V_n), \qquad (8.18)$$

where $\{\eta_i\}_{i \geq 0}$ is a standard symmetric nearest neighbor random walk on the two-dimensional lattice $\mathbb{Z}^2$, starting at x, and $\tau_{\partial V_n}$ is the first entrance time in ∂V_n. The law of Φ_n is the Gaussian distribution with density function proportional to

$$\exp\left(-\frac{1}{8}\sum_{x \sim y}(\phi_x - \phi_y)^2\right), \qquad (8.19)$$

where $x \sim y$ means x and y are neighbors in V_n (each pair counted once), with the understanding that we set $\phi_x \equiv 0$ for $x \in \partial V_n$ in the above formula. This can be seen as follows. Fix $y \in \text{int}(V_n)$, and for each $x \in \text{int}(V_n)$ let $f(x)$ and $g(x)$ denote the left- and right-hand sides of (8.18). Using (8.19) it is easy to show that for any x,

$$\mathbb{E}(\phi_x \mid (\phi_z)_{z \neq x}) = \frac{1}{4} \sum_{z \in V_n,\, z \sim x} \phi_z.$$

It follows that the function f is discrete harmonic on $\text{int}(V_n) \setminus \{y\}$, that is,

$$f(x) = \mathbb{E}\big(\mathbb{E}(\phi_x \mid (\phi_z)_{z \neq x})\phi_y\big) = \frac{1}{4} \sum_{z \in V_n,\, z \sim x} f(z).$$

Similarly, it is easy to show that $g(x)$ is discrete harmonic on $\text{int}(V_n)\setminus\{y\}$ by conditioning on η_1. Again, putting

$$\bar{\phi}_y = \mathbb{E}(\phi_y \mid (\phi_z)_{z\neq y}) = \frac{1}{4}\sum_{z\in V_n,\, z\sim y}\phi_z,$$

we have

$$f(y) - \frac{1}{4}\sum_{z\in V_n,\, z\sim y} f(z) = \mathbb{E}\big((\phi_y - \bar{\phi}_y)\phi_y\big)$$

$$= \mathbb{E}\big((\phi_y - \bar{\phi}_y)^2\big) + \mathbb{E}\big((\phi_y - \bar{\phi}_y)\bar{\phi}_y\big)$$

$$= \mathbb{E}\big(\text{Var}(\phi_y \mid (\phi_z)_{z\neq y})\big) + 0$$

$$= 1.$$

Similarly, conditioning on η_1, we can show

$$g(y) - \frac{1}{4}\sum_{z\in V_n,\, z\sim y} g(z) = 1.$$

Thus, $f - g$ is discrete harmonic on $\text{int}(V_n)$. But $f = g = 0$ on ∂V_n. Thus we must have $f = g$ everywhere on V_n, which proves that the density (8.19) indeed corresponds to the Gaussian field with covariance (8.18).

It was shown by Bolthausen et al. (2001, Lemma 1) that as $n \to \infty$, we have

$$\max_{y\in V_n}\text{Var}(\phi_y) = \frac{2}{\pi}\log n + O(1).$$

The unexpected and surprising fact, also in the same paper of Bolthausen et al. (2001, Theorem 2), is that as $n \to \infty$,

$$\mathbb{E}\Big(\max_{y\in V_n}\phi_y\Big) \sim 2\sqrt{\frac{2}{\pi}\log n},$$

exactly as if $\{\phi_y\}_{y\in V_n}$ were independent. Here $a_n \sim b_n$ means, as usual, that $\lim_{n\to\infty} a_n/b_n = 1$. In our terminology, the DGFF with zero boundary condition is extremal. Combined with Theorem 8.1, a direct consequence is the following result.

Proposition 8.6 (Proposition 10.1 in Chatterjee 2008b) *The discrete Gaussian free field on an $n \times n$ grid with zero boundary condition is superconcentrated (as $n \to \infty$), meaning that*

$$\text{Var}\Big(\max_{y\in V_n}\phi_y\Big) = o(\log n).$$

Consequently, the two dimensional DGFF is chaotic and has the multiple peaks property.

4 Application to the Discrete Gaussian Free Field

Superconcentration of the DGFF maximum under zero boundary condition was first established in the form of the above proposition in Chatterjee (2008b). Since then, there has been enormous progress on this problem. It is currently known that the true order of fluctuations is $O(1)$ as $n \to \infty$, and tight tail bounds are also known. See Bramson and Zeitouni (2011), Ding (2011), Ding and Zeitouni (2012) and Bramson et al. (2013) for details.

Chapter 9
Further Applications of Hypercontractivity

This chapter contains several applications of hypercontractivity in proving superconcentration, partly in the same vein as Talagrand's method from Chap. 5. We begin with an argument of Ledoux (2003, 2007) that gives an easy proof of the $n^{-1/6}$ fluctuations of the largest eigenvalue of a GUE matrix.

1 Superconcentration of the Largest Eigenvalue

Recall the Gaussian Unitary Ensemble (GUE) introduced in Sect. 6 of Chap. 3. Let λ_1 be the largest eigenvalue of a GUE matrix of order n. In this section we use hypercontractivity to obtain a sharp upper tail bound for λ_1, reproducing an argument of Ledoux (2003, 2007). This argument is a combination of hypercontractive estimates and properties of Hermite polynomials. We also reproduce from Ledoux (2007) a proof of the corresponding lower tail bound (which is a direct argument, not involving hypercontractivity). The two bounds combined give a proof of the superconcentration of the largest eigenvalue of a GUE matrix. Incidentally, although the fact that the largest eigenvalue has fluctuations of order $n^{-1/6}$ was proved by Tracy and Widom (1994a,b, 1996), the variance bound does not follow directly from their work.

Proposition 9.1 (Equation (4) in Ledoux 2003) *Let λ_1 be the largest eigenvalue of a standard GUE matrix of order n, defined in Sect. 6 of Chap. 3. There are constants C and c independent of n, such that for any $\epsilon \in (0, 1]$,*

$$\mathbb{P}(\lambda_1 \geq 2\sqrt{n}(1+\epsilon)) \leq C\epsilon^{-1/2} e^{-cn\epsilon^{3/2}}.$$

Proof Let γ denote the standard Gaussian measure on $\mathbb{R}$. From Sect. 1 of Chap. 6 we know that the Hermite polynomials $(H_k)_{k\geq 0}$ form an orthonormal basis of $L^2(\gamma)$.

Let $\lambda_1 \geq \cdots \geq \lambda_n$ be the eigenvalues of our GUE matrix. From the basic random matrix machinery pertaining to the GUE, one can deduce that for any bounded measurable $f : \mathbb{R} \to \mathbb{R}$,

$$\mathbb{E}\left(\sum_{i=1}^{n} f(\lambda_i)\right) = \int_{\mathbb{R}} f(x) \sum_{k=0}^{n-1} H_k^2(x)\,d\mu(x). \tag{9.1}$$

A derivation of this identity may be found in Ledoux (2007), where it is recorded as (1.13).

Let P_t be the Ornstein-Uhlenbeck semigroup with generator L. Since H_k is an eigenfunction of $-L$ with eigenvalue k, and $P_t = e^{tL}$, therefore we have the basic identity

$$P_t H_k = e^{-tk} H_k. \tag{9.2}$$

Take any $r > 1$ and choose t such that $2r = 1 + e^{2t}$. Then by (9.2) and the hypercontractive inequality (B.2) for the OU semigroup,

$$\|H_k\|_{L^{2r}(\gamma)} = e^{tk}\|P_t H_k\|_{L^{2r}(\gamma)}$$
$$\leq e^{tk}\|H_k\|_{L^2(\gamma)} = e^{tk} = (2r-1)^{k/2}. \tag{9.3}$$

By (9.1), for any $\epsilon > 0$,

$$\mathbb{P}\left(\lambda_1 \geq 2\sqrt{n}(1+\epsilon)\right) \leq \mathbb{E}\left(\sum_{i=1}^{n} 1_{\{\lambda_i \geq 2\sqrt{n}(1+\epsilon)\}}\right)$$

$$= \int_{2\sqrt{n}(1+\epsilon)}^{\infty} \sum_{k=0}^{n-1} H_k^2(x)\,d\gamma(x). \tag{9.4}$$

Applying Hölder's inequality, one gets

$$\int_{2\sqrt{n}(1+\epsilon)}^{\infty} H_k^2(x)\,d\gamma(x) \leq \gamma\left([2\sqrt{n}(1+\epsilon), \infty)\right)^{1-1/r} \|H_k\|_{L^{2r}(\gamma)}^2.$$

Applying the inequality (A.2) from Appendix A to the first factor on the right, and inequality (9.3) to the second factor, we get

$$\int_{2\sqrt{n}(1+\epsilon)}^{\infty} H_k^2(x)\,d\gamma(x) \leq e^{-2n(1+\epsilon)^2(1-1/r)}(2r-1)^k.$$

By inequality (9.4), this gives

$$\mathbb{P}\left(\lambda_1 \geq 2\sqrt{n}(1+\epsilon)\right) \leq e^{-2n(1+\epsilon)^2(1-1/r)} \sum_{k=0}^{n-1} (2r-1)^k$$

$$\leq \frac{1}{2(r-1)} e^{-2n(1+\epsilon)^2(1-1/r) + n\log(2r-1)}.$$

1 Superconcentration of the Largest Eigenvalue

The proof is completed by optimizing in r, upon taking r close to 1 and Taylor expanding the logarithm up to the third order. □

Proposition 9.2 (Equation (5.16) in Ledoux 2007) *Let λ_1 be the largest eigenvalue of a standard GUE matrix of order n. There are constants C and c independent of n, such that for any $\epsilon \in (n^{-2/3}, 1]$,*

$$\mathbb{P}\big(\lambda_1 \leq 2\sqrt{n}(1-\epsilon)\big) \leq C e^{-cn\epsilon^{3/2}}.$$

Proof From standard results in random matrix theory, it follows that

$$\mathbb{P}\big(\lambda_1 \leq 2t\sqrt{n}\big) = \det(I - K),$$

where det denotes 'determinant', I is the identity matrix of order n, and K is the $n \times n$ matrix whose (k, l)th entry is

$$\int_t^\infty H_{k-1}(x) H_{l-1}(x) d\gamma(x).$$

A proof of this well-known formula may be found in classical texts such as Mehta (1991). Note that for any unit vector $y = (y_1, \ldots, y_n) \in \mathbb{R}^n$,

$$\sum_{k,l=1}^n y_k y_l \int_t^\infty H_{k-1}(x) H_{l-1}(x) d\gamma(x) = \int_t^\infty \left(\sum_{k=1}^n y_k H_{k-1}(x) \right)^2 d\gamma(x)$$

$$\leq \int_{-\infty}^\infty \left(\sum_{k=1}^n y_k H_{k-1}(x) \right)^2 d\gamma(x)$$

$$= 1.$$

This shows that the eigenvalues $\rho_1, \ldots, \rho_n$ of K are all in the interval $[0, 1]$. Consequently,

$$\mathbb{P}\big(\lambda_1 \leq 2t\sqrt{n}\big) = \prod_{i=1}^n (1 - \rho_i) \leq e^{-\sum_{i=1}^n \rho_i}$$

$$= e^{-\operatorname{Tr}(K)} = e^{-n\mu_n([t,\infty))}, \qquad (9.5)$$

where μ_n is a probability measure on $\mathbb{R}$ defined as

$$d\mu_n(x) = \frac{1}{n} \sum_{k=0}^{n-1} H_k^2(x) d\gamma(x).$$

For every $p \geq 1$, define

$$b_{n,p} := \int_\mathbb{R} x^{2p} d\mu_n(x).$$

By the Cauchy-Schwarz inequality,

$$\left(\int_t^\infty x^{4p} d\mu(x)\right)^2 \leq \mu\big([t,\infty)\big) b_{n,4p}. \tag{9.6}$$

On the other hand

$$b_{n,2p} \leq t^{2p} b_{n,p} + 2\int_t^\infty x^{4p} d\mu_n(x),$$

which gives, for $t \in [0,1]$,

$$\int_t^\infty x^{4p} d\mu_n(x) \geq b_{n,2p} - b_{n,p}. \tag{9.7}$$

Combining (9.6) and (9.7) with (9.5) gives, for any $t \in [0,1]$ and any $p \geq 1$,

$$\mathbb{P}\big(\lambda_1 \leq 2t\sqrt{n}\big) \leq e^{-(b_{n,2p}-b_{n,p})^2/b_{n,4p}}. \tag{9.8}$$

At this point, Ledoux (2007, Proposition 5.2) proves the three-term recursion relation

$$b_{n,p} = \frac{2p-1}{2p+2} b_{n,p-1} + \frac{2p-1}{2p+2}\frac{2p-3}{2p}\frac{p(p-1)}{4n^2} b_{n,p-2},$$

which leads to upper and lower estimates for the b's, which, in turn, give an upper bound on the right-hand side of (9.8). The proof is completed by taking $t = 1 - \epsilon$ and optimizing over p. We omit the details, which may be found in Ledoux (2007, Sects. 5.1 and 5.3). □

Proposition 9.3 *Let λ_1 be the largest eigenvalue of a GUE matrix of order n. Then $\mathrm{Var}(\lambda_1) \leq Cn^{-1/3}$, where C does not depend on n. In particular, λ_1 is superconcentrated.*

Proof The bound on the variance is a direct application of Propositions 9.1 and 9.2, together with the Gaussian concentration inequality (A.5) from Appendix A to deal with the case $\epsilon > 1$. It is not difficult to prove that the derivative of λ_1 with respect to the (i,j)th matrix entry is exactly equal to $u_i u_j$, where $u = (u_1, \ldots, u_n)$ is the first eigenvector of the matrix, normalized to have Euclidean norm 1. Consequently, the norm-squared of the gradient of λ_1 with respect to the matrix entries is exactly equal to 1. This proves that λ_1 is superconcentrated. □

2 A Different Hypercontractive Tool

Talagrand's hypercontractive method can be applied only to functions of independent random variables. Although any Gaussian vector is nominally a linear function of independent Gaussian random variables, it may be difficult in practice to directly work with the underlying set of independent variables. The main result of this section gives a direct way to prove superconcentration for maxima of Gaussian random vectors with correlation. The proof uses the dynamical representation (1.4) of Chap. 1.

Theorem 9.4 *Suppose $g = (g_1, \ldots, g_n)$ is a centered Gaussian random vector with $R(i, j) := \mathrm{Cov}(g_i, g_j)$. Suppose that for each $r \geq 0$, there is a covering $\mathcal{C}(r)$ of $\{1, \ldots, n\}$ such that whenever i, j are indices with $R(i, j) \geq r$, we have $i, j \in D$ for some $D \in \mathcal{C}(r)$. Let $I := \mathrm{argmax}_i\, g_i$, and define*

$$\rho(r) := \max_{D \in \mathcal{C}(r)} \mathbb{P}(I \in D),$$

$$\mu(r) := \sum_{D \in \mathcal{C}(r)} \mathbb{P}(I \in D) = \mathbb{E}\big|\{D \in \mathcal{C}(r) : I \in D\}\big|.$$

Let $\sigma^2 := \max_i \mathrm{Var}(g_i)$. Then

$$\mathrm{Var}\Big(\max_i g_i\Big) \leq \int_0^{\sigma^2} \frac{2\mu(r)(1 - \rho(r))}{\log(1/\rho(r))}\, dr,$$

where we interpret $(1 - x)/\log(1/x) = -1$ when $x = 1$.

Proof For each $A \subseteq \{1, \ldots, n\}$ let $p(A) := \mathbb{P}(I \in A)$. Let g^t and I^t be as in Chap. 1, Sect. 2.3, and let P_t be the Ornstein-Uhlenbeck semigroup. Take any $p > 1$ and let $q = 1 + e^{2t}(p - 1)$. Let $q' = q/(q - 1)$. Then for any map $f : \mathbb{R}^n \to [0, 1]$, by the hypercontractive inequality (B.2),

$$\mathbb{E}\big(f(g)f(g^t)\big) = \mathbb{E}\big(f(g) P_t f(g)\big) \leq \|f(g)\|_{q'} \|P_t f(g)\|_q$$

$$\leq \|f(g)\|_{q'} \|f(g)\|_p$$

$$\leq \big(\mathbb{E}f(g)\big)^{\frac{1}{q'} + \frac{1}{p}}.$$

Taking $p = 1 + e^{-t}$ gives

$$\frac{1}{q'} + \frac{1}{p} = 1 - \frac{1}{1 + e^{2t}(p - 1)} + \frac{1}{p}$$

$$= 1 + \tanh(t/2).$$

Therefore,

$$\mathbb{P}\big(R(I^0, I^t) \geq r\big) \leq \sum_{D \in \mathcal{C}(r)} \mathbb{P}(I^0 \in D,\ I^t \in D)$$

$$\leq \sum_{D \in \mathcal{C}(r)} \mathbb{P}(I \in D)^{1 + \tanh(t/2)}$$

$$\leq \rho(r)^{\tanh(t/2)} \sum_{D \in \mathcal{C}(r)} \mathbb{P}(I \in D) = \mu(r)\rho(r)^{\tanh(t/2)}.$$

Thus, by Theorem 2.4,

$$\mathrm{Var}\Big(\max_i g_i\Big) = \int_0^\infty e^{-t}\mathbb{E}\big(R(I^0, I^t)\big)\, dt$$

$$\leq \int_0^\infty \int_0^{\sigma^2} e^{-t}\mathbb{P}\big(R(I^0, I^t) \geq r\big)\, dr\, dt$$

$$\leq \int_0^{\sigma^2} \int_0^\infty e^{-t}\mu(r)\rho(r)^{\tanh(t/2)}\, dt\, dr.$$

For any fixed r,

$$\int_0^\infty e^{-t}\rho(r)^{\tanh(t/2)}\, dt \leq \int_0^\infty e^{-t}\rho(r)^{1-e^{-t/2}}\, dt$$

$$\leq \int_0^\infty e^{-t/2}\rho(r)^{1-e^{-t/2}}\, dt$$

$$= \int_0^1 2\rho(r)^u\, du = \frac{2(1-\rho(r))}{-\log \rho(r)}.$$

This completes the proof. □

3 Superconcentration in Low Correlation Fields

Theorem 9.4 was used for various purposes in Chatterjee (2008b). For example, it was used to prove the following result about superconcentration of the maximum of Gaussian fields with 'low correlations'.

Proposition 9.5 *Let $g = (g_1, \ldots, g_n)$ be a centered Gaussian field with covariance matrix R. Suppose that $R(i, i) = 1$ for each i, and ϵ is a positive constant such that $R(i, j) \leq \epsilon$ for each $i \neq j$. Then*

$$\mathrm{Var}\Big(\max_i g_i\Big) \leq \frac{C}{\log n} + C\epsilon,$$

where C is a universal constant.

Proof Let $\mathcal{C}(r)$ consist of the single set $\{1, \ldots, n\}$ when $r \leq \epsilon$, and consist of the sets $\{1\}, \ldots, \{n\}$ when $r > \epsilon$. Then $\mathcal{C}(r)$ satisfies the conditions of Theorem 9.4. Also, clearly, $\mu(r) = 1$ for each r.

By increasing the value of the constant in the statement of the theorem (and recalling that $\mathrm{Var}(\max_i g_i) \leq \max_i \mathrm{Var}(g_i)$), we can assume without loss of generality that $\epsilon \leq 1/2$. Under this assumption, by the Sudakov minoration technique from Appendix A, we have

$$m := \mathbb{E}\left(\max_i g_i\right) \geq C\sqrt{\log n}.$$

Thus, by the inequalities (A.7) and (A.2) from Appendix A we have

$$\mathbb{P}(I = i) \leq \mathbb{P}(g_i \geq m/2) + \mathbb{P}\left(\max_i g_i \leq m/2\right)$$

$$\leq 2e^{-m^2/8} \leq n^{-C}.$$

In particular, $\rho(r) \leq n^{-C}$ for $r > \epsilon$. By Theorem 9.4, this completes the proof. □

4 Superconcentration in Subfields

Proposition 9.5 has the interesting consequence that maxima of subfields of Gaussian fields may have smaller variance than the maximum of the whole field. One just has to find a subfield such that the correlation between any two distinct members of the subfield is uniformly small. For example, recall the following Gaussian field introduced in Sect. 3.1 of Chap. 1. Let $g_1, \ldots, g_n$ be i.i.d. standard Gaussian random variables. Define a random function $f : \{-1, 1\}^n \to \mathbb{R}$ as

$$f(\sigma) = \sum_{i=1}^n g_i \sigma_i.$$

Then f is a Gaussian field on the hypercube $\{-1, 1\}^n$, whose maximum is attained at $\hat{\sigma} = (\hat{\sigma}_1, \ldots, \hat{\sigma}_n)$, where $\hat{\sigma}_i = \text{sign}(g_i)$. Therefore

$$\max_\sigma f(\sigma) = \sum_{i=1}^n |g_i|,$$

which has variance of order n, matching the upper bound given by the Gaussian Poincaré inequality. In other words, the maximum of the field f is not superconcentrated. On the other hand, using Proposition 9.5 one can easily prove that there are subfields whose maxima are superconcentrated, as demonstrated by the following result.

Proposition 9.6 *In the Gaussian field f defined above, there is a subset of indices $S \subseteq \{-1, 1\}^n$ such that*

$$\text{Var}\left(\max_{\sigma \in S} f(\sigma)\right) \leq C n^{2/3},$$

where C is a universal constant.

Proof Let σ and σ' be two configurations chosen uniformly at random from the hypercube. A simple application of the concentration inequality of Hoeffding (1963) shows that for any $x \geq 0$,

$$\mathbb{P}(|\sigma \cdot \sigma'| \geq x) \leq 2e^{-x^2/2n},$$

where $\sigma \cdot \sigma'$ is the usual inner product between σ and σ'. Consequently, if N is a positive integer and $\sigma^1, \ldots, \sigma^N$ are N configurations chosen independently and uniformly at random, then

$$\mathbb{P}\left(\max_{1 \leq i \neq j \leq N} |\sigma^i \cdot \sigma^j| \geq x\right) \leq 2N^2 e^{-x^2/2n}.$$

Now choose N to be the integer part of $2^{n^{1/3}}$. The above inequality proves that there exists a set of configurations $S = \{\sigma^1, \ldots, \sigma^N\}$ such that $|\sigma^i \cdot \sigma^j| \leq Cn^{2/3}$ for all $1 \leq i \neq j \leq n$, where C is a constant that does not depend on n. Since $\mathrm{Cov}(f(\sigma), f(\sigma')) = \sigma \cdot \sigma'$, an application of Proposition 9.5 completes the proof. □

Similar subfields may be constructed in other examples, such as the following.

Exercise 9.7 In the $(d+1)$-dimensional Gaussian random polymer model introduced in Sect. 1.2 of Chap. 1, prove that there is a subset $\mathcal{P}'_n$ of paths of length n, such that if E'_n is the minimum energy among all such paths, then $\mathrm{Var}(E'_n) \leq Cn^\alpha$ for some constant C that does not depend on n and some number $\alpha < 1$. What is the best possible α in your construction, depending on the dimension d?

5 Discrete Gaussian Free Field on a Torus

Recall the discrete Gaussian free field (DGFF) defined in Sect. 4 of Chap. 8. In this section we consider the free field on an $n \times n$ torus (instead of the zero boundary condition), where we can take advantage of the symmetry to prove that the variance of the maximum is $O(\log \log n)$. The main tool is Theorem 9.4. Of course, this is still not as good as the most recent results (cited in Chap. 8, Sect. 4) that give $O(1)$ bounds.

Let $\mathbb{T}_n$ be the set $\{0, \ldots, n-1\}^2$ endowed with the graph structure of a torus, that is, (a, b) and (c, d) are adjacent if $a - c \equiv \pm 1 \pmod{n}$ and $b - d \equiv \pm 1 \pmod{n}$. We wish to define a Gaussian free field on this graph. However, the graph has no natural boundary. The easiest (and perhaps the most natural) way to overcome this problem is to modify the definition (8.18) of the covariance by replacing the stopping time $\tau_{\partial V_n}$ with a random time τ that is of the same order of magnitude as $\tau_{\partial V_n}$, but is independent of all else. Specifically, we prescribe

$$\mathrm{Cov}(\phi_x, \phi_y) = \mathbb{E}_x\left(\sum_{i=0}^\tau \mathbb{I}_{\{\eta_i = y\}}\right), \tag{9.9}$$

5 Discrete Gaussian Free Field on a Torus

where $(\eta_i)_{i\geq 0}$ is a simple random walk on the torus starting at x, and τ is a random variable independent of $(\eta_i)_{i\geq 0}$, which we take to be Geometrically distributed with mean n^2. The reason for the particular choice of the Geometric distribution is that it translates into a simple modification of the density (8.19) by the introduction of a small 'mass'; the new density function turns out to be

$$\exp\left(-\frac{(1-\frac{1}{n^2})}{8}\sum_{x\sim y}(\phi_x - \phi_y)^2 - \frac{1}{2n^2}\sum_x \phi_x^2\right). \quad (9.10)$$

Here $x \sim y$ means that x and y are neighbors on the torus. To show that this density indeed corresponds to that of a centered Gaussian field with covariance (9.9), we proceed as in the case of the DGFF with zero boundary condition. Fix $y \in \mathbb{T}_n$, and let $f(x)$ and $g(x)$ be the two sides of (9.9). From (9.9) and (9.10), one can check using conditional expectations that for $x \neq y$,

$$f(x) = \frac{n^2-1}{4n^2}\sum_{z\in V_n,\, z\sim x} f(z), \qquad g(x) = \frac{n^2-1}{4n^2}\sum_{z\in V_n,\, z\sim x} g(z).$$

(The second identity holds because conditional on $\tau \geq 1$, $\tau - 1$ has the same distribution as τ. This is where we use that τ has a Geometric law.) Again, using similar computations as before, it can be checked that

$$f(y) = 1 + \frac{n^2-1}{4n^2}\sum_{z\in V_n,\, z\sim x} f(z),$$

$$g(y) = 1 + \frac{n^2-1}{4n^2}\sum_{z\in V_n,\, z\sim y} g(z).$$

Combining the last two displays, one can conclude that $|f - g|$ is a non-negative strictly subharmonic function on $\mathbb{T}_n$, which implies that it must be zero everywhere.

Although we assign a small mass in our definition of the DGFF on the torus, we can still call it a 'massless free field' in an asymptotic sense because the stopping times $\tau_{\partial V_n}$ and τ are both of order n^2 and it is not difficult to show that the covariances in the two models differ by $O(1)$.

Let us now state the main result of this section, which shows that the DGFF on the torus is superconcentrated, with an explicit bound of order $\log \log n$ on the variance of the maximum.

Theorem 9.8 (Theorem 10.2 in Chatterjee 2008b) *Let $(\phi_x)_{x\in\mathbb{T}_n}$ be the DGFF on the torus defined above. For some universal constant C, we have*

$$\mathrm{Var}\left(\max_{x\in\mathbb{T}_n}\phi_x\right) \leq C\log\log n.$$

The proof of this result is via the use of hypercontractivity, more specifically Theorem 9.4. The key advantage in the torus model is that we know that the location of the maximum is uniformly distributed. In the zero boundary situation, we had very little information about the location of the maximum.

Let us now proceed to prove Theorem 9.8. The first step is a basic observation about the simple symmetric random walk on $\mathbb{Z}$.

Lemma 9.9 *Suppose $(\alpha_i)_{i \geq 0}$ is a simple symmetric random walk on $\mathbb{Z}$, starting at 0. Then for any $k \in \mathbb{Z}$ and $i \geq 1$, we have*

$$\mathbb{P}(\alpha_i = k) \leq \frac{Ce^{-k^2/4i}}{\sqrt{i}}$$

where C is a universal constant.

Proof If $|k| > i$ or $k \not\equiv i \pmod{2}$, then $\mathbb{P}(\alpha_i = k) = 0$ and we have nothing to prove. If $|k| = i$, we have $\mathbb{P}(\alpha_i = k) = 2^{-i}$, which is consistent with the statement of the lemma. In all other cases,

$$\mathbb{P}(\alpha_i = k) = \binom{i}{\frac{i+k}{2}} 2^{-i}.$$

Using the Stirling approximation, we get

$$\mathbb{P}(\alpha_i = k) \leq \frac{C}{\sqrt{i}} \exp\left(-\frac{i+k+1}{2}\log\left(1+\frac{k}{i}\right) - \frac{i-k+1}{2}\log\left(1-\frac{k}{i}\right)\right)$$

$$= \frac{C}{\sqrt{i}} \exp\left(-iI\left(\frac{k}{i}\right) - \frac{1}{2}\log\left(1-\frac{k^2}{i^2}\right)\right),$$

where

$$I(x) = \frac{1+x}{2}\log(1+x) + \frac{1-x}{2}\log(1-x).$$

The function I has a power series expansion

$$I(x) = \sum_{p=1}^{\infty} \frac{x^{2p}}{(2p-1)2p}, \quad x \in (-1, 1).$$

In particular, $I(x) \geq x^2/2$. This inequality suffices to prove the lemma when, say, $k \leq i/2$. On the other hand, if $i/2 < k < i$ and i is so large that $\log i \leq i/8$ (which can be assumed by choosing a suitably large C), we have

$$-\frac{1}{2}\log\left(1-\frac{k^2}{i^2}\right) \leq \frac{\log i}{2} \leq \frac{i}{16} \leq \frac{k^2}{4i},$$

which implies that

$$P(\alpha_i = k) \leq \frac{2}{\sqrt{2\pi i}} \exp\left(-\frac{k^2}{2i} + \frac{k^2}{4i}\right).$$

This completes the proof. □

We are going to use random walks on $\mathbb{Z}^2$ to produce random walks on $\mathbb{T}_n$. For this purpose, we observe that there is a natural map $Q_n : \mathbb{Z}^2 \to \mathbb{T}_n$ which takes a point $(x_1, x_2) \in \mathbb{Z}^2$ to the unique point (x'_1, x'_2) in $\mathbb{T}_n$ satisfying $x_1 \equiv x'_1$ (mod n) and $x_2 \equiv x'_2$ (mod n). For any $x, y \in \mathbb{T}_n$, define the toric Euclidean distance $d_n(x, y)$ as:

$$d_n(x, y) := d\left(x, Q_n^{-1}(y)\right),$$

where $d(x, A)$ is the usual Euclidean distance of a point x from a set A. It is not difficult to verify that actually $d_n(x, y) = d(Q_n^{-1}(x), Q_n^{-1}(y))$ and therefore the definition is symmetric in x and y.

Lemma 9.10 *Let $(\eta_i)_{i\geq 0}$ be a simple symmetric random walk on $\mathbb{T}_n$. Then for any $x, y \in \mathbb{T}_n$ and any $i \geq 1$, we have*

$$\mathbb{P}_x(\eta_i = y) \leq \begin{cases} Ci^{-1} e^{-d_n(x,y)^2/4i} & \text{if } i \leq n^2, \\ Cn^{-2} & \text{if } i > n^2, \end{cases}$$

where C is a universal constant and $\mathbb{P}_x$ denotes the law of the random walk starting from x.

Proof Let $(\beta_i)_{i\geq 0}$ be a simple symmetric random walk on $\mathbb{Z}^2$. Then a random walk on the torus is easily obtained as $\eta_i = Q_n(\beta_i)$. For any $x, y \in \mathbb{T}_n$, we have

$$\mathbb{P}_x(\eta_i = y) = \sum_{z \in Q_n^{-1}(y)} \mathbb{P}_x(\beta_i = z). \tag{9.11}$$

Now, for any x and z, Lemma 9.9 shows that

$$\mathbb{P}_x(\beta_i = z) \leq \frac{C}{i} \exp\left(-\frac{d(x,z)^2}{4i}\right), \tag{9.12}$$

where $d(x, z)$ is the Euclidean distance between x and z. Now fix $x, y \in \mathbb{T}_n$, and let z be the nearest point to x in $Q_n^{-1}(y)$. Then, if $x = (x_1, x_2)$ and $z = (z_1, z_2)$, we have by (9.11) and (9.12) that

$$\mathbb{P}_x(\eta_i = y) \leq \frac{C}{i} \sum_{k_1, k_2 \in \mathbb{Z}} \exp\left(-\frac{(x_1 - z_1 + k_1 n)^2 + (x_2 - z_2 + k_2 n)^2}{4i}\right).$$

It is easy to see that $|x_j - z_j| \le n/2$ for $j = 1, 2$ (otherwise, we can choose a 'better' z_j so that z is closer to x.) Thus, for $j = 1, 2$,

$$(x_j - z_j + k_j n)^2 = (x_j - z_j)^2 + k_j^2 n^2 + 2(x_j - z_j)k_j n$$
$$\ge (x_j - z_j)^2 + |k_j|(|k_j| - 1)n^2.$$

Therefore,

$$\mathbb{P}_x(\eta_i = y) \le \frac{C}{i} e^{-d(x,z)^2/4i} \sum_{k_1, k_2 = 0}^{\infty} \exp\left(-\frac{(k_1(k_1 - 1) + k_2(k_2 - 1))n^2}{4i}\right)$$

$$\le \frac{C}{i} e^{-d(x,z)^2/4i} \left(1 + \sum_{k=1}^{\infty} e^{-k^2 n^2 / 4i} + \sum_{r,s=1}^{\infty} e^{-(r^2 + s^2)n^2 / 4i}\right).$$

Comparing the last two terms with integrals, we get

$$\mathbb{P}_x(\eta_i = y) \le \frac{C}{i} e^{-d(x,z)^2/4i} \left(1 + \int_{\mathbb{R}} e^{-u^2 n^2 / 4i} \, du + \int_{\mathbb{R}^2} e^{-(u^2 + v^2) n^2 / 4i} \, du \, dv\right)$$

$$\le \frac{C}{i} e^{-d(x,z)^2/4i} \left(1 + \frac{\sqrt{i}}{n} + \frac{i}{n^2}\right).$$

It is not difficult to verify by considering the cases $i \le n^2$ and $i > n^2$ that this completes the proof. □

Lemma 9.11 *For any $x \ne y \in \mathbb{T}_n$, we have*

$$0 \le \mathrm{Cov}(\phi_x, \phi_y) \le C \log \frac{n}{d_n(x, y)} + C$$

and $\mathrm{Var}(\phi_x) \le C \log n$, where C is a universal constant.

Proof From the representation (9.9), we see that the covariances are non-negative. Now fix two distinct points $x, y \in \mathbb{T}_n$, and let $d = d_n(x, y)$. From (9.9), we have

$$\mathrm{Cov}(\phi_x, \phi_y) = \sum_{i=0}^{\infty} \mathbb{P}_x(\eta_i = y)\mathbb{P}(\tau \ge i)$$

$$\le \sum_{i=1}^{n^2} \frac{Ce^{-d^2/4i}}{i} + \sum_{i=n^2}^{\infty} \frac{C}{n^2}\left(1 - \frac{1}{n^2}\right)^i.$$

Clearly, the second sum can be bounded by a constant that does not depend on n. For the first, note that by the inequality $e^{-x} \le x^{-1}$ that holds for $x \ge 1$, we have

$$\sum_{1 \le i \le n^2} \frac{Ce^{-d^2/4i}}{i} = \sum_{1 \le i \le d^2} \frac{Ce^{-d^2/4i}}{i} + \sum_{d^2 < i \le n^2} \frac{Ce^{-d^2/4i}}{i}$$

$$\leq \sum_{1\leq i\leq d^2} \frac{C}{d^2} + \sum_{d^2<i\leq n^2} \frac{C}{i}$$
$$\leq C + C\left(\log n^2 - \log d^2\right).$$

The bound on the variance follows similarly. This completes the proof. □

Proof of Theorem 9.8 Fix some $r \leq C\log n$, where C is the same constant as in Lemma 9.11. If $\mathrm{Cov}(\phi_x, \phi_y) \geq r$, then by Lemma 9.11,

$$d_n(x, y) \leq ne^{1-(r/C)} =: s. \tag{9.13}$$

Let A be an s-net of $\mathbb{T}_n$ for the metric d_n (i.e. a set of points that are mutually separated from each other by a distance of $> s$, and is maximal with respect to this property). Let $\mathcal{C}(r)$ be the collection of all $2s$-balls around the points of A. Then by (9.13) we see that whenever $\mathrm{Cov}(\phi_x, \phi_y) \geq r$, we must have that $x, y \in D$ for some $D \in \mathcal{C}(r)$. By symmetry, we see that the probability of the maximum being at any point $z \in \mathbb{T}_n$ is exactly n^{-2}. Therefore, the probability of the maximum being in any given $D \in \mathcal{C}(r)$ is bounded by Ks^2/n^2, where K is a universal constant. Thus, in the terminology of Theorem 9.4,

$$\rho(r) \leq \frac{Ks^2}{n^2} = K'e^{-2r/C},$$

where $K' = Ke^2$. We can assume that $K' > 1$. Again, if $x \in \mathbb{T}_n$ and $D \in \mathcal{C}(r)$ contains x, then the center of D must be at distance $\leq 2s$ from x. Now, the centers of the members of $\mathcal{C}(r)$ are separated by a distance of $\geq s$ from each other. Clearly, the maximum number of s-separated points in a $2s$-ball can be bounded by a universal constant that does not depend on n and s. It follows that the number of members of D that contain x is bounded by a universal constant. Therefore, in the notation of Theorem B.2, $\mu(r)$ is bounded by a universal constant. Using the bounds on $\rho(r)$ and $\mu(r)$ in Theorem 9.4 and the inequality $1 - x \leq -\log x$ for $x > 0$, we see that for some universal constant κ,

$$\mathrm{Var}\left(\max_x \phi_x\right) \leq C\log K' + \int_{C\log K'}^{C\log n} \frac{\kappa}{(2r/C) - \log K'}\, dr.$$

It is easy to see that the right-hand side is bounded by a constant multiple of $\log\log n$. This completes the proof. □

6 Gaussian Fields on Euclidean Spaces

Consider a stationary centered Gaussian process $(X_n)_{n\geq 0}$. If we have $\mathrm{Cov}(X_0, X_n)$ decaying to zero faster than $1/\log n$ as $n \to \infty$, then it is known at least since the work of Berman (1964) that $M_n := \max\{X_0, \ldots, X_n\}$ has fluctuations of order $(\log n)^{-1/2}$ and upon proper centering and scaling, converges to the Gumbel distribution in the limit. This result has seen considerable generalizations for one dimensional Gaussian processes, both in discrete and continuous time. Some examples are Pickands (1967, 1969a,b), and Mittal and Ylvisaker (1975). For a survey of the classical literature, we refer to the book of Leadbetter et al. (1983).

The question is considerably harder in dimensions higher than one. A large number of sophisticated results and techniques for analyzing the behavior of the maxima of higher dimensional *smooth* Gaussian fields are now known; see the excellent recent book of Adler and Taylor (2007) for a survey. Here 'smooth' usually means twice continuously differentiable. However, in the absence of smoothness, maxima of high dimensional Gaussian fields are still quite intractable. If only the expected size of the maximum is of interest, advanced techniques exist (see Talagrand 2005). The question of fluctuations is much more difficult. In fact, for general (nonsmooth) processes, even the one-dimensional work of Pickands (1967, 1969a,b) is considerably nontrivial.

One basic question one may ask is the following: what is a sufficient condition for the fluctuation of the maximum in a box of side length T to decrease like $(\log T)^{-1/2}$? In other words, when does the maximum behave as if it were the maximum of a collection of i.i.d. Gaussians, one per each unit area in the box? Classical theory (e.g. Mittal and Ylvisaker 1975) tells us that this is true for stationary one-dimensional Gaussian processes whenever the correlation between X_0 and X_T decreases at least as fast as $1/\log T$. Note that this requirement for the rate of decay is rather mild, considering that it ensures that the maximum behaves just like the maximum of independent variables.

In dimensions ≥ 2, the above question is unresolved. Here questions may also arise about maxima over subsets that are not necessarily boxes. Moreover, what if the correlation decays slower than $1/\log T$? In this section, we attempt to answer these questions. Our achievements are modest: we only have upper bounds on the variances. The issue of limiting distributions is not solved here.

Let $X = (X_u)_{u \in \mathbb{R}^d}$ be a centered Gaussian field on $\mathbb{R}^d$ with $\mathbb{E}(X_u^2) = 1$ for each u. For any Borel set $A \subseteq \mathbb{R}^d$, let

$$M(A) := \sup_{u \in A} X_u, \qquad m(A) := \mathbb{E}\big(M(A)\big).$$

For any $u \in \mathbb{R}^d$ and $r > 0$, let $B(u, r)$ denote the open ball of radius r and center u. Assume that

$$L := \sup_{u \in \mathbb{R}^d} m\big(B(u, 1)\big) < \infty. \tag{9.14}$$

Note that in particular, the above condition is satisfied when the field is stationary and continuous. Next, suppose $\phi : [0, \infty) \to \mathbb{R}$ is a decreasing function such that for all $u, v \in \mathbb{R}^d$,

$$\mathrm{Cov}(X_u, X_v) \leq \phi\big(|u - v|\big),$$

where $|u - v|$ denotes the Euclidean distance between u and v. Assume that

$$\lim_{s \to \infty} \phi(s) = 0. \tag{9.15}$$

For a set $A \subseteq \mathbb{R}^d$ and $\epsilon > 0$, let $N(A, \epsilon)$ be the maximum number of points that can be found in A such that any two them are separated by a distance of greater than ϵ (such a collection is usually called an ϵ-net of A). When $\epsilon = 1$, we simply write $N(A)$ instead of $N(A, 1)$. The following theorem is the main result of this section.

Theorem 9.12 (Theorem 11.1 in Chatterjee 2008b) *Assume (9.14) and (9.15). Then for any Borel set $A \subseteq \mathbb{R}^d$ such that $\mathrm{diam}(A) > 1$, we have*

$$\mathrm{Var}(M(A)) \leq C_1(\phi, d)\left(\phi(N(A)^{C_2(\phi, d)}) + \frac{1}{\log N(A)}\right),$$

where $C_1(\phi, d)$ and $C_2(\phi, d)$ are constants that depend only on the function ϕ and the dimension d (and not on the set A), and $N(A)$ is defined above.

Remarks The first observation is that if $\phi(s)$ decreases at least as fast as $1/\log s$, then the above result clearly shows that

$$\mathrm{Var}(M(A)) \leq \frac{C(\phi, d)}{\log N(A)},$$

where $C(\phi, d)$ is some constant that depends only on ϕ and d. In particular, it gives a broad generalization of the classical results about fluctuations in one dimension, developed in Berman (1964), Pickands (1967, 1969a,b) and Mittal and Ylvisaker (1975). An additional observation is that the first term in the bound can dominate the second only if $\phi(s)$ decreases slower than $1/\log s$.

Before we embark on the proof of Theorem 9.12, we need some simple upper and lower bounds on the expected value of $M(A)$.

Lemma 9.13 *Under (9.14) and (9.15), for any Borel set $A \subseteq \mathbb{R}^d$ such that $N(A) \geq 2$, we have*

$$c_1(\phi, d)\sqrt{\log N(A)} \leq m(A) \leq c_2(\phi, d)\sqrt{\log N(A)},$$

where $c_1(\phi, d)$ and $c_2(\phi, d)$ are positive constants that depend only on the function ϕ and the dimension d.

Proof The upper bound follows easily from a combination of assumption (9.14), the tail bound from Proposition A.7 for the maximum of the field in unit balls, and an argument similar to the proof of Lemma A.1.

For the lower bound, first let $s > 1$ be a number such that $\phi(s) < 1/2$. Such an s exists by the assumption that $\lim_{s \to \infty} \phi(s) = 0$. Next, let B be a 1-net of A and D

be an s-net of B. For each $x \in D$, the s-ball around x can contain at most k points of B, where k is a fixed number that depends only on s and the dimension d. Thus,

$$|D| \geq \frac{|B|}{k} = \frac{N(A)}{k}.$$

Since $\phi(s) < 1/2$ and $\mathbb{E}(X_u^2) = 1$ for each u, by the Sudakov minoration technique (Lemma A.3 from Appendix A) we have

$$m(A) \geq m(D) \geq C\sqrt{\log |D|} \geq c_1(\phi, d)\sqrt{\log N(A)},$$

where D is a universal constant and $c_1(\phi, d)$ is a constant depending only on the function ϕ and the dimension d. □

Proof of Theorem 9.12 Let c_1 and c_2 be the constants from Lemma 9.13. Put

$$s := N(A)^{\frac{1}{8}(c_1/c_2)^2},$$

and assume that $N(A)$ is so large that $s > 2$. Let $r = \phi(s)$. Take any maximal s-net of A, and let $\mathcal{C}(r)$ be the set of $2s$-balls around the points in the net. It is easy to verify using the definition of s and the decreasing nature of ϕ, that $\mathcal{C}(r)$ is a covering of A satisfying the conditions of Theorem 9.4. Now take any $D \in \mathcal{C}(r)$. Since D is a $2s$-ball and $s > 2$, by Lemma 9.13 we have

$$m(D) \leq c_2\sqrt{\log 2s} \leq c_2\sqrt{2\log s} = \frac{c_1}{2}\sqrt{\log N(A)}.$$

Also, by Lemma 9.13, we have $m(A) \geq c_1\sqrt{\log N(A)}$. Thus, using the notation of Theorem 9.4, we have by Proposition A.7 that

$$p(D) \leq \mathbb{P}\left(M(D) \geq m(D) + \frac{m(A) - m(D)}{2}\right)$$
$$+ \mathbb{P}\left(M(A) \leq m(A) - \frac{m(A) - m(D)}{2}\right)$$
$$\leq 2\exp\left(-\frac{(m(A) - m(D))^2}{8}\right)$$
$$\leq 2\exp\left(-\frac{c_1^2 \log N(A)}{32}\right).$$

Since the above bound does not depend on D, it serves as a bound on $\rho(r)$. Let us now get a bound on $\mu(r)$. Take any $u \in A$. Then the center of any $D \in \mathcal{C}(r)$ that contains u is a point in $B(u, 2s)$. The centers are mutually separated by a distance of more than s. Hence, the number of $D \in \mathcal{C}(r)$ that can contain u is bounded by $N(B(u, 2s), s)$, which, by scaling symmetry, is equal to $N(B(0, 2), 1)$. Since $\mu(r)$ is the expected number of elements of $\mathcal{C}(r)$ that contain the maximizer of X in A, we have that $\mu(r) \leq c_3$, where c_3 is a constant that depends only on the dimension d. Combining the bounds, we see that whenever $N(A) \geq c_4$, we have

$$\frac{\mu(r)}{|\log \rho(r)|} \leq \frac{c_5}{\log N(A)},$$

where c_4 and c_5 are constants depending only on ϕ and d. Note that we have this bound only for one specific value of r defined above. Now, if $r' > r$, and we define $\mathcal{C}(r')$ the same way as we defined $\mathcal{C}(r)$, then clearly $\mathcal{C}(r')$ would also be a cover of A satisfying the requirements of Theorem 9.4, and we would have $\rho(r') \leq \rho(r)$. Noting this, and the fact that $|(1-x)/\log x| \leq 1$ for all $x \in (0,1)$, we have by Theorem 9.4 that

$$\mathrm{Var}(M(A)) \leq c_6\left(r + \frac{1}{\log N(A)}\right),$$

for some constant c_6 depending only on ϕ and d. Of course this holds only if $N(A) \geq c_4$, but this condition can now be dropped by increasing the value of c_6, since we always have $\mathrm{Var}(M(A)) \leq 1$. Plugging in the value of r, the proof is done. □

Chapter 10
The Interpolation Method for Proving Chaos

The purpose of this chapter is to prove a generalization of Theorem 1.11 of Chap. 1. Recall that a slightly weaker version of this theorem has already been proved in Chap. 6 using the spectral method. The proof of the stronger version requires a technique that I call 'the interpolation method'.

1 A General Theorem

Let $g = (g_1, \ldots, g_n)$ be a centered Gaussian field with covariance matrix R. Let g' be an independent copy of g, and let $g^t := e^{-t}g + \sqrt{1 - e^{-2t}}\,g'$, as in Chap. 1, Sect. 2.3. Let $[n]$ denote the set $\{1, \ldots, n\}$.

Fix a real number $\beta \geq 0$. For each $t, s \geq 0$, define a probability measure $G_{t,s}$ on $[n] \times [n]$ that assigns mass

$$\frac{e^{\beta g_i^t + \beta g_j^s}}{\sum_{k,l} e^{\beta g_k^t + \beta g_l^s}}$$

to the point (i, j). The average of a function $h : [n] \times [n] \to \mathbb{R}$ under the measure $G_{t,s}$ will be denoted by $\langle h \rangle_{t,s}$, that is,

$$\langle h \rangle_{t,s} := \frac{\sum_{i,j} h(i,j) e^{\beta g_i^t + \beta g_j^s}}{\sum_{i,j} e^{\beta g_i^t + \beta g_j^s}}.$$

We will consider the covariance matrix R as a function on $[n] \times [n]$. Alternatively, it will also be considered as a square matrix.

Theorem 10.1 (Theorem 3.1 in Chatterjee 2009) *Suppose that for all i, j, $R(i, j) \geq 0$. For each i, let*

$$v_i := \mathbb{E}\left(\frac{e^{\beta g_i}}{\sum_j e^{\beta g_j}}\right).$$

Let $\phi(x) = \sum_{k=0}^{\infty} c_k x^k$ be any convergent power series on $[0, \infty)$ all of whose coefficients are non-negative. Then for each $t \geq 0$,

$$\mathbb{E}\langle \phi \circ R\rangle_{0,t} \leq \inf_{s \geq t}\left(\mathbb{E}\langle \phi \circ R\rangle_{0,0}\right)^{1-t/s} \left(\sum_{i,j} \phi(R(i,j)) e^{2\beta^2 e^{-s} R(i,j)} v_i v_j\right)^{t/s}.$$

Moreover, $\mathbb{E}\langle \phi \circ R\rangle_{0,t}$ is a decreasing function of t.

We will see later in this chapter how to use the above theorem for proving strong inequalities for chaos in the Sherrington-Kirkpatrick, and in particular, to prove Theorem 1.11 from Chap. 1.

In the following, $C_b^{\infty}(\mathbb{R}^n)$ will denote the set of all infinitely differentiable real-valued functions on $\mathbb{R}^n$ with bounded derivatives of all orders.

Let us first extend the definition of g^t to negative t. Let g'' be another independent copy of g that is also independent of g', and for each $t \geq 0$, let

$$g^{-t} := e^{-t} g + \sqrt{1 - e^{-2t}}\, g''.$$

Lemma 10.2 *For any $f \in C_b^{\infty}(\mathbb{R}^n)$, we have*

$$\frac{d}{dt}\mathbb{E}(f(g^{-t}) f(g^t)) = -2e^{-2t} \sum_{i,j} R(i,j) \mathbb{E}(\partial_i f(g^{-t}) \partial_j f(g^t)),$$

where $\partial_i f$ is the partial derivative of f in the ith coordinate.

Proof For each $t \geq 0$, define

$$h^t := \sqrt{1 - e^{-2t}}\, g - e^{-t} g', \qquad h^{-t} := \sqrt{1 - e^{-2t}}\, g - e^{-t} g''.$$

A simple computation gives

$$\frac{d}{dt}\mathbb{E}(f(g^{-t}) f(g^t))$$
$$= -\frac{e^{-t}}{\sqrt{1 - e^{-2t}}} \mathbb{E}\big((h^{-t} \cdot \nabla f(g^{-t})) f(g^t) + (h^t \cdot \nabla f(g^t)) f(g^{-t})\big)$$
$$= -\frac{2e^{-t}}{\sqrt{1 - e^{-2t}}} \mathbb{E}\big((h^{-t} \cdot \nabla f(g^{-t})) f(g^t)\big).$$

(Note that issues like moving derivatives inside expectations are easily taken care of due to the assumption that $f \in C_b^{\infty}$.) One can verify by computing covariances that h^{-t} and the pair (g^{-t}, g') are independent. Moreover,

$$g^t = e^{-2t} g^{-t} + e^{-t}\sqrt{1 - e^{-2t}}\, h^{-t} + \sqrt{1 - e^{-2t}}\, g'.$$

1 A General Theorem

So for any i, Gaussian integration by parts (see Appendix A) gives

$$\mathbb{E}(h_i^{-t}\partial_i f(g^{-t})f(g^t)) = e^{-t}\sqrt{1-e^{-2t}}\sum_j R(i,j)\mathbb{E}(\partial_i f(g^{-t})\partial_j f(g^t)).$$

The proof is completed by combining the last two steps. □

Lemma 10.3 *Let $\mathcal{F}$ be the class of all functions u on $[0, \infty)$ that can be expressed as*

$$u(t) = \sum_{i=1}^{m} e^{-c_i t}\mathbb{E}(f_i(g^{-t})f_i(g^t))$$

for some non-negative integer m and non-negative real numbers $c_1, c_2, \ldots, c_m$, and functions $f_1, \ldots, f_m$ in $C_b^\infty(\mathbb{R}^S)$. For any $u \in \mathcal{F}$, there is a probability measure μ on $[0, \infty)$ such that for each $t \geq 0$,

$$u(t) = u(0)\int_{[0,\infty)} e^{-xt}\,d\mu(x).$$

In particular, for any $0 < t \leq s$,

$$u(t) \leq u(0)^{1-t/s} u(s)^{t/s}.$$

Proof Note that any $u \in \mathcal{F}$ must necessarily be a non-negative function, since g^{-t} and g^t are independent and identically distributed conditional on g, which gives

$$\mathbb{E}(f(g^{-t})f(g^t)) = \mathbb{E}((\mathbb{E}(f(g^t)\mid g))^2).$$

Now, if $u(0) = 0$, then $u(t) = 0$ for all t, and there is nothing to prove. So let us assume $u(0) > 0$.

Since R is a positive semidefinite matrix, there is a square matrix C such that $R = C^T C$. Thus, given a function f, if we define

$$w_i := \sum_j C_{ij}\partial_j f,$$

then by Lemma 10.2 we have

$$\frac{d}{dt}\mathbb{E}(f(g^{-t})f(g^t)) = -2e^{-2t}\sum_i \mathbb{E}(w_i(g^{-t})w_i(g^t)).$$

From this observation and the definition of $\mathcal{F}$, it follows easily that if $u \in \mathcal{F}$, then $-u' \in \mathcal{F}$. Proceeding by induction, we see that for any k, $(-1)^k u^{(k)}$ is a non-negative function (where $u^{(k)}$ denotes the kth derivative of u). Such functions on $[0, \infty)$ are called 'completely monotone'. The most important property of completely monotone functions (see e.g. Feller 1971, Vol. II, Sect. XIII.4) is that any such function u can be represented as the Laplace transform of a positive Borel measure μ on $[0, \infty)$, that is,

$$u(t) = \int_{[0,\infty)} e^{-xt} dv(x).$$

Moreover, $u(0) = v(\mathbb{R})$. By taking $\mu(dx) = u(0)^{-1} v(dx)$, this proves the first assertion of the theorem. For the second, note that by Hölder's inequality, we have that for any $0 < t \leq s$,

$$\int_{\mathbb{R}} e^{-xt} d\mu(x) \leq \left(\int_{\mathbb{R}} e^{-xs} d\mu(x) \right)^{t/s} = \left(u(s)/u(0) \right)^{t/s}.$$

This completes the proof. □

The next lemma is obtained by a variant of the Gaussian interpolation methods for analyzing mean field spin glasses at high temperatures. It is similar to R. Latała's unpublished proof of the replica symmetric solution of the SK model (see Talagrand 2011).

Lemma 10.4 *Let ϕ and v_i be as in Theorem 10.1. Then for each $t \geq 0$,*

$$\mathbb{E}\langle \phi \circ R \rangle_{-t,t} \leq \sum_{i,j} \phi(R(i,j)) e^{2\beta^2 e^{-2t} R(i,j)} v_i v_j.$$

Proof For each i, define a function $p_i : \mathbb{R}^n \to \mathbb{R}$ as

$$p_i(x) := \frac{e^{\beta x_i}}{\sum_j e^{\beta x_j}}.$$

Note that

$$\partial_j p_i = \beta(p_i \delta_{ij} - p_i p_j),$$

where $\delta_{ij} = 1$ if $i = j$ and 0 otherwise. Since p_i is bounded, this proves in particular that $p_i \in C_b^{\infty}(\mathbb{R}^n)$.

Take any non-negative integer r. Since R is a positive semidefinite matrix, so is $R^{(r)} := (R(i,j)^r)_{1 \leq i,j \leq n}$. (To see this, just note that $g^1, \ldots, g^r$ are independent copies of g, then $\text{Cov}(g_i^1 \cdots g_i^r, g_j^1 \cdots g_j^r) = R(i,j)^r$.) Therefore there exists a matrix $C^{(r)} = (C_{ij}^{(r)})$ such that $R^{(r)} = (C^{(r)})^T C^{(r)}$. Define the functions

$$w_i := \sum_j C_{ij}^{(r)} p_j, \quad i \in [n].$$

1 A General Theorem

In the following we will denote $p_i(g^s)$ and $w_i(g^s)$ by p_i^s and h_i^s respectively, for all $s \in \mathbb{R}$. Let

$$f_r(t) := \mathbb{E}\left(\sum_{i,j} R(i,j)^r p_i^{-t} p_j^t\right) = \mathbb{E}\left(\sum_i w_i^{-t} w_i^t\right).$$

By Lemma 10.2 we get

$$f_r'(t) = -2e^{-2t} \sum_i \sum_{k,l} R(k,l) \mathbb{E}(\partial_k w_i^{-t} \partial_l w_i^t)$$

$$= -2\beta^2 e^{-2t} \sum_{i,k,l} R(k,l) \mathbb{E}\left(\left(C_{ik}^{(r)} p_k^{-t} - \sum_j C_{ij}^{(r)} p_j^{-t} p_k^{-t}\right)\right.$$

$$\left. \times \left(C_{il}^{(r)} p_l^t - \sum_j C_{ij}^{(r)} p_j^t p_l^t\right)\right)$$

$$= -2\beta^2 e^{-2t} \mathbb{E}\left(\sum_{k,l} R(k,l)^{r+1} p_k^{-t} p_l^t - \sum_{j,k,l} R(k,l) R(j,l)^r p_j^{-t} p_k^{-t} p_l^t \right.$$

$$\left. - \sum_{j,k,l} R(k,l) R(k,j)^r p_k^{-t} p_j^t p_l^t + \sum_{j,k,l,m} R(k,l) R(j,m)^r p_j^{-t} p_k^{-t} p_m^t p_l^t\right).$$

Our objective is to get a lower bound for $f_r'(t)$. For this, we can delete the two middle terms in the above expression because they contribute a positive amount. For the fourth term, note that by Hölder's inequality,

$$\left(\sum_{k,l} R(k,l) p_k^{-t} p_l^t\right)\left(\sum_{j,m} R(j,m)^r p_j^{-t} p_m^t\right)$$

$$\leq \left(\sum_{k,l} R(k,l)^{r+1} p_k^{-t} p_l^t\right)^{\frac{1}{r+1}} \left(\sum_{j,m} R(j,m)^{r+1} p_j^{-t} p_m^t\right)^{\frac{r}{r+1}}$$

$$= \sum_{k,l} R(k,l)^{r+1} p_k^{-t} p_l^t.$$

Thus, by Lemma 10.3 and the above inequalities, we have

$$0 \geq f_r'(t) \geq -4\beta^2 e^{-2t} f_{r+1}(t). \tag{10.1}$$

Now let $u_r(x) := f_r(-\log \sqrt{x})$ for $0 < x < 1$. Then

$$u_r'(x) = -\frac{f_r'(-\log \sqrt{x})}{2x}.$$

The inequality (10.1) becomes

$$0 \leq u_r'(x) \leq 2\beta^2 u_{r+1}(x). \tag{10.2}$$

Fix $0 < x < 1$, $r \geq 1$. For each $m \geq 1$, let

$$T_m := \int_0^x \int_0^{x_1} \cdots \int_0^{x_{m-1}} (2\beta^2)^{m-1} u'_{r+m-1}(x_m) dx_m dx_{m-1} \cdots dx_1.$$

Using (10.2), we see that

$$0 \leq T_m \leq \int_0^x \int_0^{x_1} \cdots \int_0^{x_{m-1}} (2\beta^2)^m u_{r+m}(x_m) dx_m dx_{m-1} \cdots dx_1$$

$$= \int_0^x \cdots \int_0^{x_{m-1}} (2\beta^2)^m \left(u_{r+m}(0) + \int_0^{x_m} u'_{r+m}(x_{m+1}) dx_{m+1} \right) dx_m \cdots dx_1$$

$$= \frac{(2\beta^2)^m u_{r+m}(0) x^m}{m!} + T_{m+1}.$$

Inductively, this implies that for any $m \geq 1$,

$$u_r(x) = u_r(0) + T_1 \leq \sum_{l=0}^{m} \frac{u_{r+l}(0)(2\beta^2 x)^l}{l!} + T_{m+1}.$$

Again, for any $m \geq 2$,

$$0 \leq T_m \leq \int_0^x \int_0^{x_1} \cdots \int_0^{x_{m-1}} (2\beta^2)^m u_{r+m}(x_m) dx_m dx_{m-1} \cdots dx_1$$

$$\leq \frac{M^{r+m}(2\beta^2 x)^m}{m!},$$

where $M = \max_{i,j} R(i, j)$. Thus, $\lim_{m \to \infty} T_m = 0$. Finally, observe that p_i^∞ and $p_i^{-\infty}$ are independent. This implies that for any m,

$$u_m(0) = f_m(\infty) = \sum_{i,j} R(i, j)^m v_i v_j.$$

Combining, we conclude that

$$u_r(x) \leq \sum_{l=0}^{\infty} \frac{u_{r+l}(0)(2\beta^2 x)^l}{l!} = \sum_{i,j} R(i, j)^r e^{2\beta^2 x R(i,j)} v_i v_j.$$

The result now follows easily by taking $x = e^{-2t}$ and summing over r, using the fact that ϕ has non-negative coefficients in its power series. □

Proof of Theorem 10.1 Let p_i^t be as in the proof of Lemma 10.4. As noted in the proof of Lemma 10.4, the matrix $(R(i, j)^r)$ is positive semidefinite for every non-negative integer r. Since ϕ has non-negative coefficients in its power series, it follows that the matrix $\Phi := (\phi(R(i,j)))_{1\leq i,j \leq n}$ is also positive semidefinite. Let $C = (C_{ij})$ be a matrix such that $\Phi = C^T C$. Then

$$\langle \phi \circ R \rangle_{-t,t} = \sum_{i,j} \phi(R(i,j)) p_i^{-t} p_j^t = \sum_i \left(\sum_j C_{ij} p_j^{-t} \right) \left(\sum_j C_{ij} p_j^t \right).$$

Therefore, the function

$$u(t) := \mathbb{E}\langle \phi \circ R \rangle_{-t,t}$$

belongs to the class $\mathcal{F}$ of Lemma 10.3. The proof is now finished by using Lemma 10.3 and Lemma 10.4, and the observation that $u(t/2) = \mathbb{E}\langle \phi \circ R \rangle_{0,t}$ (since $(g^{-t/2}, g^{t/2})$ has the same law as (g^0, g^t)). The claim that $u(t)$ is a decreasing function of t is automatic because $u' \leq 0$. $\square$

2 Application to the Sherrington-Kirkpatrick Model

We are now ready to give a proof of Theorem 1.11 of Chap. 1 using Theorem 10.1. In fact, we will prove a slightly more general result below, which also covers the case of p-spin models for even p, as well as further generalizations.

Let n be a positive integer and suppose $(H_n(\sigma))_{\sigma \in \{-1,1\}^n}$ is a centered Gaussian random vector with

$$\text{Cov}(H_n(\sigma), H_n(\sigma')) = n\xi\left(\frac{\sigma \cdot \sigma'}{n}\right),$$

where ξ is some function on $[-1, 1]$ that does not depend on n and $\sigma \cdot \sigma'$ is the inner product between σ and σ'. Let us fix $\beta \geq 0$. The Hamiltonian H_n defines a Gibbs measure on $\{-1, 1\}^n$ by putting mass proportional to $e^{-\beta H_n(\sigma)}$ at each configuration σ. This class of models was considered by Talagrand (2006) in his proof of the generalized Parisi formula. For the SK model, $\xi(x) = x^2/2$, while for the p-spin models, $\xi(x) = x^p/p!$. (We refer to Talagrand 2003, Chap. 6 for the definition of the p-spin models and related discussions.)

Let H_n' be an independent copy of H_n, and for each $t \geq 0$, let

$$H_n^t := e^{-t} H_n + \sqrt{1 - e^{-2t}} H_n'.$$

Given a function h on $\{-1, 1\}^n \times \{-1, 1\}^n$, we define the average $\langle h(\sigma, \sigma') \rangle_{t,s}$ as the average with respect to the product of the Gibbs measures defined by H_n^t and H_n^s, that is,

$$\langle h(\sigma, \sigma') \rangle_{t,s} := \frac{\sum_{\sigma,\sigma'} h(\sigma, \sigma') e^{-\beta H_N^t(\sigma) - \beta H_N^s(\sigma')}}{\sum_{\sigma,\sigma'} e^{-\beta H_N^t(\sigma) - \beta H_N^s(\sigma')}}.$$

The following result establishes the presence of chaos in this class of models under some restrictions on ξ. It is easy to see that the result covers all p-spin models for even p, and in particular, the original SK model.

Theorem 10.5 (Generalization of Theorem 1.11 of Chap. 1) *Consider the model defined above. Suppose that ξ is non-negative, $\xi(1) = 1$ and that there is a constant c such that $\xi(x) \leq cx^2$ for all $x \in [-1, 1]$. Then there is a constant C depending only on c such that for all $t \geq 0$ and $\beta > 1$, and any positive integer k,*

$$\mathbb{E}\left\langle \xi\left(\frac{\sigma \cdot \sigma'}{n}\right)^k \right\rangle_{0,t} \leq (Ck)^k n^{-k \min\{1, t/C \log(1+C\beta)\}}.$$

Proof By symmetry, it is easy to see that for each σ,

$$\mathbb{E}\left(\frac{e^{-\beta H_n(\sigma)}}{\sum_{\sigma'} e^{-\beta H_n(\sigma')}}\right) = 2^{-n}.$$

Again, it follows from elementary combinatorial arguments that there are positive constants γ and C that do not depend on n, such that for any positive integer k and any n,

$$2^{-2n} \sum_{\sigma,\sigma'} \frac{(\sigma \cdot \sigma')^{2k}}{n^k} \exp\left(\frac{\gamma(\sigma \cdot \sigma')^2}{n}\right) \leq (Ck)^k.$$

Choosing s so that $2\beta^2 c e^{-s} = \min\{\gamma, 2\beta^2 c\}$, and $\phi(x) = x^k/n^k$, we see from Theorem 10.1 (and the assumption that $\xi(x) \leq cx^2$) that for any $0 \leq t \leq s$

$$\mathbb{E}\left\langle \xi\left(\frac{\sigma \cdot \sigma'}{n}\right)^k \right\rangle_{0,t}$$

$$\leq \left(\mathbb{E}\left\langle \xi\left(\frac{\sigma \cdot \sigma'}{n}\right)^k \right\rangle_{0,0}\right)^{1-t/s} \left(c^k 2^{-2n} \sum_{\sigma,\sigma'} \frac{(\sigma \cdot \sigma')^{2k}}{n^{2k}} \exp\left(\frac{\gamma(\sigma \cdot \sigma')^2}{n}\right)\right)^{t/s}$$

$$\leq (Ck)^k n^{-kt/s},$$

where C is a constant that does not depend on n. This proves the result for $t \leq s$. For $t \geq s$ we use the last assertion of Theorem 10.1 to conclude that $\mathbb{E}\langle \xi(\sigma \cdot \sigma'/n)^k\rangle_{0,t}$ is decreasing in t. Finally, observe that

$$s = \begin{cases} 0 & \text{if } 2\beta^2 c \leq \gamma, \\ \log(2\beta^2 c/\gamma) & \text{if } 2\beta^2 c > \gamma \end{cases}$$

$$\leq C \log(1 + C\beta)$$

for some constant C that depends only on c and γ. □

3 Sharpness of the Interpolation Method

The Random Energy Model (REM), introduced by Derrida (1980, 1981), is possibly the simplest model of a spin glass. The state space is $\{-1, 1\}^n$ as usual, but here the energies of states $\{-H_n(\sigma)\}_{\sigma\in\{-1,1\}^n}$ are chosen to be i.i.d. Gaussian random variables with mean zero and variance n. We show that Theorem 10.1 gives a sharp result in the low temperature regime ($\beta > 2\sqrt{\log 2}$) of this model. We follow the notation of Theorem 10.1.

Proposition 10.6 *Suppose σ^1 is drawn from the original Gibbs measure of the REM and σ^2 from the Gibbs measure perturbed up to time t, in the sense of Sect. 2.3 of Chap. 1. If $\beta > 2\sqrt{\log 2}$, there are positive constants $C(\beta)$ and $c(\beta)$ depending only on β such that for all n and t,*

$$c(\beta)e^{-C(\beta)n\min\{1,t\}} \leq \mathbb{E}\langle 1_{\{\sigma^1=\sigma^2\}}\rangle_{0,t} \leq C(\beta)e^{-c(\beta)n\min\{1,t\}}.$$

Proof In the notation of Theorem 10.1, we have $R(\sigma,\sigma') = 0$ if $\sigma \neq \sigma'$, and $R(\sigma,\sigma') = n$ if $\sigma = \sigma'$. Also, clearly, $\nu_\sigma = 2^{-n}$ for each σ. Suppose σ^1 is drawn from the original Gibbs measure and σ^2 from the Gibbs measure perturbed continuously up to time t. Taking $\phi(x) = x/n$ in Theorem 10.1, we get

$$\mathbb{E}\langle 1_{\{\sigma^1=\sigma^2\}}\rangle_{0,t} \leq \inf_{s\geq t}\left(2^{-n}e^{2\beta^2 e^{-s}n}\right)^{t/s}.$$

Now choose s so large that $2\beta^2 e^{-s} \leq \frac{1}{2}\log 2$. The above inequality shows that for $t \leq s$,

$$\mathbb{E}\langle 1_{\{\sigma^1=\sigma^2\}}\rangle_{0,t} \leq 2^{-nt/2s}, \tag{10.3}$$

and for $t > s$,

$$\mathbb{E}\langle 1_{\{\sigma^1=\sigma^2\}}\rangle_{0,t} \leq 2^{-n}e^{2\beta^2 e^{-t}n}. \tag{10.4}$$

A simple computation using the covariance lemma (Lemma 2.1 of Chap. 2) now gives $\mathrm{Var}(F_n(\beta)) \leq C(\beta)$, where $C(\beta)$ is a constant depending only on β. Now suppose $\beta > 2\sqrt{\log 2}$. Let $H'_n(\sigma) = H_n(\sigma) + na_n$, where a_n solves

$$na_n^2 = \log\left(\frac{2^n}{\sqrt{n}}\right).$$

Let $(w_\alpha^n)_{1\leq\alpha\leq 2^n}$ denote the numbers $\exp(-\beta H'_n(\sigma))$ when enumerated in non-increasing order. It follows from arguments in Sect. 1.2 of Talagrand (2003) that this point process converges in distribution, as $n \to \infty$, to a Poisson point process $(w_\alpha)_{\alpha\geq 1}$ with intensity x^{-m-1} on $[0,\infty)$, where $m = 2\sqrt{\log 2}/\beta$. It is not difficult to extend this argument to show that

$$\lim_{n\to\infty} \mathrm{Var}\left(\log\sum_{\alpha=1}^{2^n} w_\alpha^n\right) = \mathrm{Var}\left(\log\sum_{\alpha=1}^{\infty} w_\alpha\right) > 0.$$

We skip the details, which are somewhat tedious. (Here $\beta > 2\sqrt{\log 2}$ is required to ensure that the infinite sum $\sum_1^\infty w_\alpha$ converges almost surely.)

However, $\text{Var}(\log \sum w_\alpha^n) = \text{Var}(\beta F_n(\beta))$. Thus, there is a positive constant $c(\beta)$ depending only on β such that for any n, $\text{Var}(F_n(\beta)) \geq c(\beta)$. We can now proceed as in the proof of Theorem 11.4 of Chap. 11 to prove that for some positive constant $c(\beta)$ depending only on β, we have that for any n and t,

$$\mathbb{E}\langle 1_{\{\sigma^1 = \sigma^2\}}\rangle_{0,t} \geq c(\beta)e^{-nt/c(\beta)}. \tag{10.5}$$

However, we also have by Theorem 10.1 that $\mathbb{E}\langle 1_{\{\sigma^1 = \sigma^2\}}\rangle_{0,t}$ is a decreasing function of t, and hence

$$\mathbb{E}\langle 1_{\{\sigma^1 = \sigma^2\}}\rangle_{0,t} \geq \mathbb{E}\langle 1_{\{\sigma^1 = \sigma^2\}}\rangle_{0,\infty} = 2^{-n}.$$

Combined with (10.3), (10.4) and (10.5), this completes the proof. □

Chapter 11
Variance Lower Bounds

Until now, we have only discussed how to get upper bounds on variances. Sometimes it may be important to obtain lower bounds, usually to show that some function is *not* superconcentrated. A very useful tool is the Plancherel identity (6.2) for the Gaussian measure from Chap. 6. Since each term in the identity is positive, one may obtain a lower bound on the variance by including a finite number of terms and ignoring the rest.

1 Some General Tools

In this section we present two lemmas that make precise the method outlined in the opening paragraph of this chapter.

Lemma 11.1 *Suppose $f : \mathbb{R}^n \to \mathbb{R}$ is an absolutely continuous function such that there is a version of its gradient ∇f that is bounded on bounded sets. Let g be a standard Gaussian random vector in $\mathbb{R}^n$, and suppose $\mathbb{E}|f(g)|^2$ and $\mathbb{E}|\nabla f(g)|^2$ are both finite. Then*

$$\mathrm{Var}(f(g)) \geq \frac{1}{2} \sum_{i=1}^n \left(\mathbb{E}(g_i \partial_i f(g))\right)^2 \geq \frac{1}{2n} \left(\mathbb{E}(g \cdot \nabla f(g))\right)^2,$$

where $x \cdot y$ denotes the usual inner product on $\mathbb{R}^n$.

Proof First assume that $f \in C_b^\infty$. The identity (6.2) from Chap. 6 implies that

$$\mathrm{Var}(f(g)) \geq \frac{1}{2} \sum_{i=1}^n \left(\mathbb{E}(\partial_i^2 f(g))\right)^2.$$

Integration by parts gives

$$\mathbb{E}(\partial_i^2 f(g)) = \mathbb{E}((g_i^2 - 1) f(g)).$$

Thus, for any C_b^∞ function f,

$$\mathrm{Var}(f(g)) \geq \frac{1}{2}\sum_{i=1}^{n}\left(\mathbb{E}\left((g_i^2-1)f(g)\right)\right)^2. \tag{11.1}$$

Let us now show that the above inequality holds for any bounded Lipschitz function f. For each $t > 0$ and $x \in \mathbb{R}^n$, define

$$f_t(x) := \mathbb{E}\big(f(x+tg)\big).$$

Then we can write

$$f_t(x) = \int_{\mathbb{R}^n} f(x+ty) \frac{e^{-\frac{1}{2}|y|^2}}{(2\pi)^{n/2}} dy$$

$$= \int_{\mathbb{R}^n} t^{-n} f(z) \frac{e^{-\frac{1}{2t^2}|z-x|^2}}{(2\pi)^{n/2}} dz.$$

Since f is a bounded function, it is clear from the above representation that $f_t \in C_b^\infty$ for any $t > 0$, and hence (11.1) holds for f_t. Again, since f is Lipschitz,

$$|f_t(x) - f(x)| \leq Lt\mathbb{E}|g|,$$

where L is the Lipschitz constant of f. This shows that we can take $t \to 0$ and obtain (11.1) for f.

Next, we want to show (11.1) whenever f is absolutely continuous and square-integrable under the Gaussian measure, and the gradient of f is bounded on bounded sets. Take any such f. Let $h : \mathbb{R}^n \to [0,1]$ be a Lipschitz function that equals 1 in the ball of radius 1 centered at the origin, and vanishes outside the ball of radius 2. For each $n \geq 1$, define

$$f_n(x) := f(x)h(n^{-1}x).$$

Then note that each f_n is bounded and Lipschitz (with possibly increasing Lipschitz constants). Thus, (11.1) holds for each f_n. Since $|f_n| \leq |f|$ everywhere, and $f_n \to f$ pointwise, and f is square-integrable under the Gaussian measure, it follows that we can take $n \to \infty$ and get (11.1) for f.

Finally, we wish to show that if ∇f is square-integrable under the Gaussian measure, we have

$$\mathbb{E}\big((g_i^2-1)f(g)\big) = \mathbb{E}\big(g_i \partial_i f(g)\big).$$

(Note that f is almost surely an absolutely continuous function of g_i if we fix $(g_j)_{j \neq i}$. This follows from Fubini's theorem.) The above identity follows from the univariate identity

$$\mathbb{E}\big(Z\phi(Z)\big) = \mathbb{E}\big(\phi'(Z)\big)$$

that holds when Z is a standard Gaussian random variable and ϕ is any absolutely continuous function such that $\mathbb{E}|\phi(Z)|$, $\mathbb{E}|Z\phi(Z)|$ and $\mathbb{E}|\phi'(Z)|$ are all finite. The identity is just integration by parts when ϕ is absolutely continuous and vanishes outside a bounded set. In the general case, let $\phi_n(x) = \phi(x)h(x/n)$, where $h : \mathbb{R} \to [0, 1]$ is a Lipschitz function that is 1 on $[-1, 1]$ and vanishes outside $[-2, 2]$. Then the above identity holds for each ϕ_n, and we can pass to the limit using the dominated convergence theorem. (Actually, it can be shown that the finiteness of $\mathbb{E}|\phi'(Z)|$ suffices.) As a last step, we observe that by the Cauchy-Schwarz inequality,

$$\sum_{i=1}^{n}\left(\mathbb{E}(g_i\partial_i f(g))\right)^2 \geq \frac{1}{n}\left(\sum_{i=1}^{n}\mathbb{E}(g_i\partial_i f(g))\right)^2.$$

This completes the proof. □

Let g be as in the previous lemma. Let A be a finite subset of $\mathbb{R}^n$. Consider the function

$$f_\beta(x) := \frac{1}{\beta}\log\sum_{y\in A}e^{\beta y\cdot x}.$$

Lemma 11.2 *For any $\beta > 0$ we have*

$$\mathrm{Var}(f_\beta(g)) \geq \sup_{0\leq\beta'\leq\beta}\frac{\beta'^2}{2n}\left(\sum_{i=1}^{n}\mathbb{E}\left(\frac{\sum_{y\in A}y_i^2 e^{\beta' y\cdot x}}{\sum_{y\in A}e^{\beta' y\cdot x}} - \left(\frac{\sum_{y\in A}y_i e^{\beta' y\cdot x}}{\sum_{y\in A}e^{\beta' y\cdot x}}\right)^2\right)\right)^2.$$

Proof Note that

$$\partial_i f_\beta(x) = \frac{\sum_{y\in A}y_i e^{\beta y\cdot x}}{\sum_{y\in A}e^{\beta y\cdot x}},$$

and therefore

$$x\cdot\nabla f_\beta(x) = \frac{\sum_{y\in A}(y\cdot x)e^{\beta y\cdot x}}{\sum_{y\in A}e^{\beta y\cdot x}} = \frac{\partial}{\partial\beta}\log\sum_{y\in A}e^{\beta y\cdot x}.$$

Now, it is easy to verify that $\log\sum e^{\beta y\cdot x}$ is a convex function of β, and hence for each x, $x\cdot\nabla f_\beta(x)$ is an increasing function of β. Thus, $\mathbb{E}(g\cdot\nabla f_\beta(g))$ is also an increasing function of β. Moreover,

$$\mathbb{E}(g\cdot\nabla f_0(g)) = \frac{1}{|A|}\sum_{y\in A}\mathbb{E}(y\cdot g) = 0,$$

and therefore $\mathbb{E}(g \cdot \nabla f_\beta(g)) \geq 0$ for all $\beta > 0$. Finally note that by integration by parts,

$$\mathbb{E}(g \cdot \nabla f_\beta(g)) = \sum_{i=1}^n \mathbb{E}(\partial_i^2 f_\beta(g))$$

$$= \beta \left(\frac{\sum_{y \in A} y_i^2 e^{\beta y \cdot x}}{\sum_{y \in A} e^{\beta y \cdot x}} - \left(\frac{\sum_{y \in A} y_i e^{\beta y \cdot x}}{\sum_{y \in A} e^{\beta y \cdot x}} \right)^2 \right).$$

Combined with Lemma 11.1, this completes the proof. □

2 Application to the Edwards-Anderson Model

As an application of the method outlined above, one can prove that the free energy in the so-called Edwards-Anderson model of spin glasses (Edwards and Anderson 1975) on a low degree graph (e.g. $\mathbb{Z}^d$) is not superconcentrated. This model is defined as follows.

Let $G = (V, E)$ be an undirected graph. The Edwards-Anderson spin glass on G is defined through the Hamiltonian

$$H(\sigma) := - \sum_{(i,j) \in E} g_{ij} \sigma_i \sigma_j, \quad \sigma \in \{-1, 1\}^V, \tag{11.2}$$

where (g_{ij}) is again a collection of i.i.d. random variables, often taken to be Gaussian. The SK model corresponds to the case of the complete graph, with an additional normalization by $\sqrt{n}$.

For a survey of the (few) rigorous and non-rigorous results available for the Edwards-Anderson model, we refer to Newman and Stein (2007).

The following result says that the free energy is not superconcentrated in the EA model on bounded degree graphs. This generalizes a well-known result of Wehr and Aizenman (1990), who proved the analogous result on square lattices. The relative advantage of our approach is that it does not use the structure of the graph, whereas the Wehr-Aizenman proof depends heavily on properties of the lattice.

Theorem 11.3 (Theorem 1.8 in Chatterjee 2009) *Let $F(\beta)$ denote the free energy in the Edwards-Anderson model on a graph G, defined as*

$$F(\beta) := -\frac{1}{\beta} \log \sum_{\sigma \in \{-1,1\}^E} e^{-\beta H(\sigma)}. \tag{11.3}$$

Let d be the maximum degree of G. Then for any β, including $\beta = \infty$ (where the free energy is just the energy of the ground state), we have

$$\operatorname{Var} F(\beta) \geq \frac{9|E|}{32} \min\left\{ \beta^2, \frac{1}{4d^2} \right\}.$$

Proof Let $\langle \cdot \rangle_\beta$ denote the average with respect to the Gibbs measure at inverse temperature β. First, we will work with $\beta < \infty$. By Lemma 11.2, with $n = |E|$, $g = (g_{ij})_{(i,j) \in E}$ and $A = \{(\sigma_i \sigma_j)_{(i,j) \in E} : \sigma \in \{-1, 1\}^V\}$, we get

$$\text{Var}(F(\beta)) \geq \sup_{0 \leq \beta' \leq \beta} \frac{\beta'^2}{2|E|} \left(|E| - \sum_{(i,j) \in E} \mathbb{E} \langle \sigma_i \sigma_j \rangle_{\beta'}^2 \right)^2.$$

Now, under the Gibbs measure at inverse temperature β', the conditional expectation of σ_i given the rest of the spins is $\tanh(\beta' \sum_{j \in N(i)} g_{ij} \sigma_j)$, where $N(i)$ is the neighborhood of i in the graph G. Using this fact and the inequality $|\tanh x| \leq |x|$, we get

$$\mathbb{E} \langle \sigma_i \sigma_j \rangle_{\beta'}^2 = \mathbb{E} \left\langle \tanh\left(\beta' \sum_{k \in N(i)} g_{ik} \sigma_k \right) \sigma_j \right\rangle^2$$

$$\leq \beta'^2 \mathbb{E} \left(\sum_{k \in N(i)} |g_{ik}| \right)^2$$

$$\leq \beta'^2 d \sum_{k \in N(i)} \mathbb{E} |g_{ik}|^2 \leq \beta'^2 d^2.$$

Thus,

$$\text{Var}(F(\beta)) \geq \frac{|E|}{2} \sup_{0 \leq \beta' \leq \min\{\beta, 1/d\}} \beta'^2 (1 - \beta'^2 d^2)^2.$$

Taking $\beta' = \min\{\beta, 1/2d\}$, we get

$$\text{Var}\, F(\beta) \geq \frac{9|E|}{32} \min\left\{ \beta^2, \frac{1}{4d^2} \right\}.$$

Finally, to prove the lower bound for $\beta = \infty$, just note that $F(\beta) \to F(\infty)$ almost surely, and the quantities are all bounded, so we can apply the dominated convergence theorem to get convergence of the variance. This completes the proof of Theorem 11.3. □

3 Chaos in the Edwards-Anderson Model

Unlike the SK model, there are two kinds of overlap in the Edwards-Anderson model. The 'site overlap' is the usual overlap defined in (1.3) of Chap. 1. The 'bond overlap' between two states σ^1 and σ^2, on the other hand, is defined as

$$Q_{1,2} := \frac{1}{|E|} \sum_{(i,j) \in E} \sigma_i^1 \sigma_j^1 \sigma_i^2 \sigma_j^2. \tag{11.4}$$

Below, we show that the bond overlap in the Edwards-Anderson model is not chaotic with respect to small fluctuations of the couplings at any temperature. This does not say anything about the site overlap; the site overlap in the Edwards-Anderson model may well be chaotic with respect to small fluctuations of the couplings, as predicted in Fisher and Huse (1986) and Bray and Moore (1987).

Theorem 11.4 (Theorem 1.6 in Chatterjee 2009) *Suppose that the Edwards-Anderson Hamiltonian* (11.2) *on a graph G is perturbed up to time* $t \geq 0$, *according to the definition of perturbation in Sect. 2.2 of Chap. 1. Let* σ^1 *be chosen from the original Gibbs measure at inverse temperature* β *and* σ^2 *is chosen from the perturbed measure. Let the bond overlap* $Q_{1,2}^{(t)}$ *be defined as in* (11.4). *Let*

$$q := \min\left\{\beta^2, \frac{1}{4d^2}\right\},$$

where d is the maximum degree of G. Then

$$\mathbb{E}(Q_{1,2}^{(t)}) \geq Cqe^{-t/Cq},$$

where C is a positive universal constant. Moreover, the result holds for $\beta = \infty$ *also, with the interpretation that the Gibbs measure at* $\beta = \infty$ *is just the uniform distribution on the set of ground states.*

Proof First, suppose that $0 \leq \beta < \infty$. Let $F(\beta)$ denote the free energy of the Edwards-Anderson model, as defined in (11.3). A simple application of the covariance lemma (Lemma 2.1 of Chap. 2) shows that

$$\mathrm{Var}(F(\beta)) = |E| \int_0^\infty e^{-t} \mathbb{E}(Q_{1,2}^{(t)}) dt.$$

By Lemma 10.3 of Chap. 10, there is a non-negative random variable U such that for all t,

$$\mathbb{E}(Q_{1,2}^{(t)}) = \mathbb{E}(Q_{1,2}^{(0)})\mathbb{E}(e^{-tU}).$$

Combined with the previous display, this gives

$$\mathrm{Var}(F(\beta)) = |E|\mathbb{E}(Q_{1,2}^{(0)})\mathbb{E}((1+U)^{-1}), \tag{11.5}$$

and consequently

$$\mathbb{E}(Q_{1,2}^{(t)}) = \frac{\mathrm{Var}(F(\beta))}{|E|\mathbb{E}((1+U)^{-1})}\mathbb{E}(e^{-tU}). \tag{11.6}$$

Let $v := \mathbb{E}((1+U)^{-1})$. Note that

$$\mathbb{E}(e^{-tU}) = \int_0^1 \mathbb{P}(e^{-tU} \geq y) dy$$

$$= \int_0^1 \mathbb{P}\big((1+U)^{-1} \geq (1-t^{-1}\log y)^{-1}\big)dy.$$

Now, for any $\epsilon > 0$, we have

$$\mathbb{E}\big((1+U)^{-1}\big) \leq \epsilon + \mathbb{P}\big((1+U)^{-1} \geq \epsilon\big).$$

Thus, if $\epsilon \leq v/2$, then

$$\mathbb{P}\big((1+U)^{-1} \geq \epsilon\big) \geq \frac{v}{2}.$$

Now $(1 - t^{-1}\log y)^{-1} \leq v/2$ if and only if $y \leq e^{-t(2-v)/v}$. Combining the steps gives

$$\mathbb{E}(e^{-tU}) \geq \int_0^{e^{-t(2-v)/v}} \frac{v}{2} dy = \frac{v}{2} e^{-t(2-v)/v}.$$

Therefore by (11.6),

$$\mathbb{E}(Q_{1,2}^{(t)}) = \frac{\text{Var}(F(\beta))}{|E|v} \mathbb{E}(e^{-tU}) \geq \frac{\text{Var}(F(\beta))}{2|E|} e^{-t(2-v)/v}.$$

By Theorem 11.3, $\text{Var}(F(\beta)) \geq C|E|q$. By (11.5), $v \geq \text{Var}(F(\beta))/|E| \geq Cq$. This completes the proof when $\beta < \infty$.

Next, note that as $\beta \to \infty$, the Gibbs measure at inverse temperature β converges weakly to the uniform distribution on the set of ground states. The same holds for the perturbed Gibbs measure. Thus,

$$\lim_{\beta \to \infty} \langle Q_{1,2}^{(t)} \rangle_\beta = \langle Q_{1,2}^{(t)} \rangle_\infty \quad \text{a.s.},$$

where $\langle Q_{1,2}^{(t)} \rangle_\beta$ denotes the Gibbs average at inverse temperature β. Since all quantities are bounded by 1, we can take expectations on both sides and apply the dominated convergence theorem. □

Theorem 11.4 establishes that the bond overlap does not become zero for any amount of perturbation. In light of this, it is surprising that the bond overlap exhibits a sort of 'quenched chaos', as demonstrated by the following theorem.

Theorem 11.5 (Theorem 1.7 in Chatterjee 2009) *Fix $t > 0$ and let $Q_{1,2}$ be as in Theorem 11.4. Then*

$$\mathbb{E}\big((Q_{1,2} - \langle Q_{1,2} \rangle)^2\big) \leq \frac{2}{\beta e^{-t/2}\sqrt{t|E|}}.$$

Proof Let $g = (g_{ij})_{(i,j) \in E}$, and let g', g'' be independent copies of g. For each t, let

$$g^t := e^{-t}g + \sqrt{1 - e^{-2t}}\, g', \qquad g^{-t} := e^{-t}g + \sqrt{1 - e^{-2t}}\, g''.$$

For each $t \in \mathbb{R}$, let σ^t denote a configuration drawn from the Gibbs measure defined by the disorder g^t. For $t \ne s$, we assume that σ^t and σ^s are independent given g, g', g''. Define

$$\phi(t) := \frac{1}{|E|} \sum_{(i,j)\in E} \mathbb{E}(\langle \sigma_i^t \sigma_j^t \rangle \langle \sigma_i^{-t} \sigma_j^{-t} \rangle).$$

By Lemma 10.3 of Chap. 10, it follows that ϕ is a completely monotone function on $[0, \infty)$. Also, ϕ is bounded by 1. Thus, for any $t > 0$,

$$|\phi'(t)| \le \frac{\phi(0) - \phi(t)}{t} \le \frac{1}{t}. \tag{11.7}$$

Again, if we let

$$e_{ijkl}^t := \langle \sigma_i^t \sigma_j^t \sigma_k^t \sigma_l^t \rangle - \langle \sigma_i^t \sigma_j^t \rangle \langle \sigma_k^t \sigma_l^t \rangle,$$

then by Lemma 10.2 of Chap. 10,

$$\phi'(t) = -\frac{2e^{-2t}\beta^2}{|E|} \sum_{(i,j)\in E, (k,l)\in E} \mathbb{E}(e_{ijkl}^t e_{ijkl}^{-t}).$$

Now fix t, and let $\bar{e}_{ijkl} := \mathbb{E}(e_{ijkl}^t \mid g)$. Then $\mathbb{E}(e_{ijkl}^t e_{ijkl}^{-t}) = \mathbb{E}(\bar{e}_{ijkl}^2)$ and so by (11.7),

$$\sum_{(i,j)\in E, (k,l)\in E} \mathbb{E}(\bar{e}_{ijkl}^2) \le \frac{|E|}{2t e^{-2t}\beta^2}. \tag{11.8}$$

Now let

$$u_{ijkl}^t := \langle \sigma_i^t \sigma_j^t \sigma_k^t \sigma_l^t \rangle, \quad v_{ijkl}^t := \langle \sigma_i^t \sigma_j^t \rangle \langle \sigma_k^t \sigma_l^t \rangle,$$

and define $\bar{u}_{ijkl} := \mathbb{E}(u_{ijkl}^t \mid g)$ and $\bar{v}_{ijkl} := \mathbb{E}(v_{ijkl} \mid g)$. Then $|\bar{u}_{ijkl}|$ and $|\bar{v}_{ijkl}|$ are both uniformly bounded by 1, and so

$$\sum_{(i,j)\in E, (k,l)\in E} \mathbb{E}(u_{ijkl}^t u_{ijkl}^{-t} - v_{ijkl}^t v_{ijkl}^{-t}) = \sum_{(i,j)\in E, (k,l)\in E} \mathbb{E}(\bar{u}_{ijkl}^2 - \bar{v}_{ijkl}^2)$$

$$\le 2 \sum_{(i,j)\in E, (k,l)\in E} \mathbb{E}|\bar{u}_{ijkl} - \bar{v}_{ijkl}|.$$

Since $\bar{u}_{ijkl} - \bar{v}_{ijkl} = \bar{e}_{ijkl}$, an application of the Cauchy-Schwarz inequality and (11.8) to the above bound gives

$$\sum_{(i,j)\in E, (k,l)\in E} \mathbb{E}(u_{ijkl}^t u_{ijkl}^{-t} - v_{ijkl}^t v_{ijkl}^{-t}) \le \frac{2|E|^{3/2}}{e^{-t}\beta\sqrt{2t}}.$$

3 Chaos in the Edwards-Anderson Model

To complete the proof, note that

$$\frac{1}{|E|^2} \sum_{(i,j)\in E, (k,l)\in E} \left(u^t_{ijkl} u^{-t}_{ijkl} - v^t_{ijkl} v^{-t}_{ijkl}\right) = \langle (Q_{\sigma^t,\sigma^{-t}} - \langle Q_{\sigma^t,\sigma^{-t}}\rangle)^2 \rangle,$$

where $Q_{\sigma^t,\sigma^{-t}}$ is the bond overlap between σ^t and σ^{-t}. □

Chapter 12
Dimensions of Level Sets

Given a centered Gaussian field $g = (g_1, \ldots, g_n)$, the goal of this chapter is to prove some results about the size of level sets, i.e. sets like

$$A_\alpha := \{i : g_i \geq \alpha\sqrt{2\log n}\}.$$

The results of this chapter are not directly related to superconcentration; they arise as interesting byproducts of the analysis of extremal fields in Chap. 8.

Drawing analogy with the definition of Hausdorff dimensions of subsets of the real line, we may define the *dimension* of the level set A_α as

$$\dim A_\alpha := \frac{\log |A_\alpha|}{\log n}.$$

Let us now see what this gives in the i.i.d. case. Suppose $g_1, \ldots, g_n$ are i.i.d. standard Gaussian random variables. Then if n is large, it is highly likely that $\dim A_\alpha \approx f(\alpha)$, where

$$f(\alpha) := \begin{cases} 1 & \text{if } \alpha \leq 0, \\ 1 - \alpha^2 & \text{if } 0 < \alpha < 1, \\ 0 & \text{if } \alpha \geq 1. \end{cases} \tag{12.1}$$

To see this, simply note that in the i.i.d. case, $|A_\alpha|$ is a Binomial(n, p) random variable, where $p = \mathbb{P}(g_1 > \alpha\sqrt{2\log n})$. When $0 < \alpha < 1$, the pair of inequalities (A.1) from Appendix A show that $p \sim n^{-\alpha^2 + o(1)}$ as $n \to \infty$. Therefore, for $0 < \alpha < 1$, $|A_\alpha| = n^{1-\alpha^2+o(1)}$ with high probability, and thus $\dim A_\alpha \approx 1 - \alpha^2$. The cases $\alpha \leq 0$ and $\alpha \geq 1$ follow by monotonicity of the dimension in α.

1 Level Sets of Extremal Fields

The following result shows that for extremal Gaussian fields (see Chap. 8), the dimension of the level set A_α is approximately $f(\alpha)$, just as if the coordinates were independent. This theorem may be used, for example, to re-derive results about the dimensions of level sets of the discrete Gaussian free field obtained by Daviaud (2006) and Hu et al. (2010).

Theorem 12.1 *Let $g = (g_1, \ldots, g_n)$ be a Gaussian random vector whose components have mean 0 and variance 1, but may be correlated. For any α let A_α be the level set $\{i : g_i \geq \alpha\sqrt{2\log n}\}$. Let $\epsilon := \mathbb{E}(\max_i g_i)/\sqrt{2\log n}$. Then for each $\alpha \in (0, 1)$, and each $\delta \in (100\epsilon, (1-\alpha)/100)$,*

$$\mathbb{P}(|\dim A_\alpha - (1-\alpha^2)| > \delta) \leq \frac{5n^{-\delta^2/16900}}{1 - n^{-\delta}}.$$

Proof Fix $\alpha \in (0, 1)$. Fix δ such that

$$100\epsilon < \delta < \frac{1-\alpha}{100}. \tag{12.2}$$

Let $\gamma := \alpha + \delta$. Note that by (12.2), $\gamma \in (0, 1)$. Let g' be an independent copy of the vector g, and let

$$h := \gamma g + \sqrt{1 - \gamma^2} g'. \tag{12.3}$$

Then h has the same distribution as g and g'. Let

$$w := \sqrt{1 - \gamma^2} g - \gamma g'. \tag{12.4}$$

Again, w has the same distribution as h, g and g'. Moreover, an easy verification of correlations shows that w and h are independent. Lastly, note that

$$g = \gamma h + \sqrt{1 - \gamma^2} w. \tag{12.5}$$

Define $I := \operatorname{argmax}_i g_i$ and $B := \{i : \alpha\sqrt{2\log n} \leq h_i \leq (\alpha + 2\delta)\sqrt{2\log n}\}$. Let E denote the event

$$\left\{|g_I - \sqrt{2\log n}| \leq \frac{\delta}{9}\sqrt{2\log n},\ |g'_I| \leq \frac{\delta}{10}\sqrt{2\log n}\right\}.$$

By the independence of g and g', the ϵ-extremality of g, the inequalities (A.2), (A.7) and (12.2),

$$\mathbb{P}(E) \geq 1 - 4n^{-\delta^2/100}.$$

Now, if E happens, then

$$h_I = \gamma g_I + \sqrt{1-\gamma^2} g'_I$$
$$\geq \left((\alpha + \delta)\left(1 - \frac{\delta}{9}\right) - \frac{\delta}{10}\right)\sqrt{2\log n} \geq \alpha\sqrt{2\log n}$$

1 Level Sets of Extremal Fields

and similarly

$$h_I \leq \left((\alpha+\delta)\left(1+\frac{\delta}{9}\right)+\frac{\delta}{10}\right)\sqrt{2\log n} \leq (\alpha+2\delta)\sqrt{2\log n}.$$

Thus, if E happens, then $I \in B$. Therefore,

$$\mathbb{P}(I \in B) \geq 1 - 4n^{-\delta^2/100}. \tag{12.6}$$

Let $\mathbb{P}^h$ denote the conditional probability given h. Let $\eta = \sqrt{\alpha^2 + 12\delta}$. Let F denote the event

$$\max_{i \in B} w_i \leq \sqrt{2(1-\eta^2)\log n},$$

where the maximum is interpreted as $-\infty$ if B is empty. Note that by (12.2), $\eta \in (0, 1)$. If $|B| \leq n^{1-\eta^2-\delta}$, then by the inequality (A.2) from Appendix A and the independence of w and h,

$$\mathbb{P}^h(F^c) \leq \sum_{i \in B} \mathbb{P}^h\left(w_i \geq \sqrt{2(1-\eta^2)\log n}\right)$$

$$\leq |B|n^{-(1-\eta^2)} \leq n^{-\delta},$$

and therefore $\mathbb{P}^h(F) \geq 1 - n^{-\delta}$. If F happens, then by (12.5) and the definition of B,

$$\max_{i \in B} g_i \leq \gamma(\alpha+2\delta)\sqrt{2\log n} + \sqrt{(1-\gamma^2)(1-\eta^2)2\log n}.$$

By the inequality $\sqrt{1-x} \leq 1 - \frac{x}{2}$,

$$\gamma(\alpha+2\delta) + \sqrt{(1-\gamma^2)(1-\eta^2)}$$

$$\leq \alpha^2 + 3\alpha\delta + 2\delta^2 + \sqrt{(1-\alpha^2)(1-\alpha^2-12\delta)}$$

$$= \alpha^2 + 5\delta + (1-\alpha^2)\sqrt{1 - \frac{12\delta}{1-\alpha^2}}$$

$$\leq 1 + 5\delta - 6\delta = 1 - \delta.$$

Therefore, if $|B| \leq n^{1-\eta^2-\delta}$, then

$$\mathbb{P}^h\left(\max_{i \in B} g_i \leq (1-\delta)\sqrt{2\log n}\right) \geq 1 - n^{-\delta}.$$

Consequently, if we set $p := \mathbb{P}(|B| \leq n^{1-\eta^2-\delta})$, then

$$\mathbb{P}\left(\max_{i \in B} g_i \leq (1-\delta)\sqrt{2\log n}\right) \geq p(1-n^{-\delta}).$$

Combined with (12.6), this gives

$$\mathbb{P}\left(\max_i g_i \leq (1-\delta)\sqrt{2\log n}\right) \geq \mathbb{P}\left(I \in B, \ \max_{i \in B} g_i \leq (1-\delta)\sqrt{2\log n}\right)$$

$$\geq p(1 - n^{-\delta}) - 4n^{-\delta^2/100}.$$

But by the ϵ-extremality of g and the inequalities (12.2) and (A.7),

$$\mathbb{P}\left(\max_i g_i \leq (1-\delta)\sqrt{2\log n}\right) \leq n^{-(99\delta/100)^2}.$$

Combining the last two displays, we get $p \leq 5n^{-\delta^2/100}/(1-n^{-\delta})$. Thus,

$$\mathbb{P}\left(|A_\alpha| \leq n^{1-\alpha^2-13\delta}\right) \leq \frac{5n^{-\delta^2/100}}{1-n^{-\delta}}.$$

By Markov's inequality and (A.2),

$$\mathbb{P}\left(|A_\alpha| \geq n^{1-\alpha^2+\delta}\right) \leq n^{-(1-\alpha^2+\delta)}\mathbb{E}|A_\alpha| \leq n^{-\delta}.$$

Replacing δ by $\delta/13$ finishes the proof. $\square$

2 Induced Dimension

Given a Gaussian field $g = (g_1, \ldots, g_n)$ where each coordinate has mean 0 and variance 1, and a set $A \subseteq \{1, \ldots, n\}$, define the *dimension of A induced by g* as

$$\dim_g(A) := \left(\frac{\mathbb{E}(\max_{i \in A} g_i)}{\mathbb{E}(\max_i g_i)}\right)^2.$$

(The maximum in the denominator is maximum over all i.) Note that $\dim_g$, like dim, is bounded between 0 and 1. The induced dimension of singleton sets is zero, and that of $\{1, \ldots, n\}$ is 1. If $A \subseteq B$, then $\dim_g(A) \leq \dim_g(B)$. The induced dimension can be related to the usual dimension by a pair of inequalities.

Proposition 12.2 *For each i and $\delta > 0$, let $B(i, \delta) := \{j : \mathbb{E}(g_i g_j) > 1 - \delta\}$. Define*

$$\theta := \frac{(\mathbb{E}(\max_i g_i))^2}{2\log n}.$$

Then for any non-empty $A \subseteq \{1, \ldots, n\}$,

$$\dim(A) \leq \inf_{\delta \in (0,1)} \left(\max_{1 \leq i \leq n} \dim B(i, \delta) + C\theta\delta^{-1} \dim_g(A)\right),$$

where C is a universal constant. Conversely, for any A,

$$\dim_g(A) \leq \theta^{-1} \dim(A).$$

2 Induced Dimension

Proof First, note that by the inequality (A.3) from Appendix A,
$$\mathbb{E}\left(\max_{i \in A} g_i\right) \leq \sqrt{2 \log |A|}.$$

Therefore for any $A \subseteq \{1, \ldots, n\}$,
$$\dim_g(A) \leq \frac{2 \log |A|}{(\mathbb{E}(\max_i g_i))^2} \leq \frac{2 \log n}{(\mathbb{E}(\max_i g_i))^2} \dim(A).$$

Next, let $V(\delta) := \max_i |B(i, \delta)|$ and for each $A \subseteq \{1, \ldots, n\}$, let
$$m(A) := \mathbb{E}\left(\max_{i \in A} g_i\right).$$

For any non-empty A and any $\delta > 0$, let $N(A, \delta)$ be the maximum number of points that can be found in A such that for any two points i and j, $\mathbb{E}(g_i g_j) \leq 1 - \delta$. Let $B \subseteq A$ be such a collection. Let $(\xi_i)_{i \in B}$ be another collection of Gaussian random variables with $\mathbb{E}(\xi_i) = 0$, $\mathbb{E}(\xi_i^2) = 1$ for each i, and $\mathbb{E}(\xi_i \xi_j) = 1 - \delta$ for each $i \neq j$. Then by Slepian's lemma (see Appendix A),
$$m(A) \geq m(B) \geq \mathbb{E}\left(\max_{i \in B} \xi_i\right).$$

Now, $(\xi_i)_{i \in B}$ can be represented as
$$\xi_i = \sqrt{1 - \delta}\, \eta + \sqrt{\delta}\, \eta_i,$$
where η and $(\eta_i)_{i \in B}$ are i.i.d. standard Gaussian random variables. Therefore
$$\mathbb{E}\left(\max_{i \in B} \xi_i\right) = \sqrt{\delta}\, \mathbb{E}\left(\max_{i \in B} \eta_i\right).$$

Since η_i are i.i.d. standard Gaussians,
$$\mathbb{E}\left(\max_{i \in B} \eta_i\right) \geq c\sqrt{\log |B|},$$
where c is a positive universal constant. Combining the last three steps, we get the inequality
$$m(A) \geq c\sqrt{\delta \log N(A, \delta)}.$$

Since $|A| \leq N(A, \delta) V(\delta)$, the result follows. □

The advantage of working with the induced dimension is that the induced dimensions of level sets can be exactly computed in the large n limit.

Theorem 12.3 *Let $g = (g_1, \ldots, g_n)$ be a Gaussian random vector whose components have mean zero and variance 1. Let $m := \mathbb{E}(\max_i g_i)$. For each $\alpha \in (0, 1)$, let*

B_α be the level set $\{i : g_i \geq \alpha m\}$. There exists a universal constant C_0 such that for any $\alpha \in (0, 1)$ and any $x \in (0, \sqrt{1-\alpha^2})$,

$$\mathbb{P}\big(\big|\dim_g(B_\alpha) - (1-\alpha^2)\big| \geq C_0 x\big) \leq \frac{4}{x} e^{-(1-\alpha^2)x^2 m^2/64}.$$

Proof Fix $0 < \alpha < \gamma < 1$ and let $\delta = (\gamma - \alpha)/2$. Let g', h, w and I be defined as in the proof of Theorem 12.1. Let $m := \mathbb{E}(\max_i g_i)$. Let F denote the event

$$\{|g_I - m| \leq \delta m, \ |g'_I| \leq \delta m\}.$$

By the independence of g and g', and the inequalities (A.2) and (A.7) from Appendix A,

$$\mathbb{P}(F) \geq 1 - 4e^{-\delta^2 m^2/2}. \tag{12.7}$$

Now, if F happens, then by (12.3),

$$|h_I - \gamma m| \leq 2\delta m.$$

In particular, $h_I \geq \alpha m$. Again, if F happens, then by (12.4),

$$\big|w_I - \sqrt{1-\gamma^2} m\big| \leq 2\delta m.$$

Let $A := \{i : h_i \geq \alpha m\}$. The previous two displays imply that if F happens, then

$$\max_{i \in A} w_i \geq \big(\sqrt{1-\gamma^2} - 2\delta\big) m.$$

Lastly by (12.5), if F happens, then

$$\max_{i \in A} w_i \leq \frac{\max_i g_i - \gamma \min_{i \in A} h_i}{\sqrt{1-\gamma^2}}$$

$$\leq \frac{(1 + \delta - \gamma\alpha) m}{\sqrt{1-\gamma^2}}.$$

The last two inequalities show that if F happens, then

$$\Big|\max_{i \in A} w_i - \sqrt{1-\alpha^2} m\Big| \leq \frac{C_0 \delta m}{\sqrt{1-\alpha^2}},$$

where C_0 is some universal constant. Let $\mathbb{E}^h$ denote the conditional expectation given h. By the above inequality,

$$\Big|\mathbb{E}^h\big(\max_{i \in A} w_i\big) - \sqrt{1-\alpha^2} m\Big| \leq \mathbb{E}^h \Big|\max_{i \in A} w_i - \sqrt{1-\alpha^2} m\Big|$$

$$\leq \frac{C_0 \delta m}{\sqrt{1-\alpha^2}} + \mathbb{E}^h\Big(\Big|\max_{i \in A} w_i - \sqrt{1-\alpha^2} m\Big|; F^c\Big).$$

By (A.7), (12.7) and the assumption that $m \geq 1$, we have

$$\mathbb{E}\left(\left|\max_i w_i - \sqrt{1-\alpha^2}m\right|; F^c\right) \leq \left(\mathbb{E}\left(\left(\max_i w_i\right)^2\right)\mathbb{P}(F^c)\right)^{1/2} + m\mathbb{P}(F^c)$$

$$\leq C_0 m e^{-\delta^2 m^2/4},$$

provided that C_0 is chosen large enough. Thus by Markov's inequality,

$$\mathbb{P}\left(\left|\mathbb{E}^h\left(\max_{i \in A} w_i\right) - \sqrt{1-\alpha^2}m\right| \geq \frac{2C_0 \delta m}{\sqrt{1-\alpha^2}}\right) \leq \frac{\sqrt{1-\alpha^2}}{\delta} e^{-\delta^2 m^2/4}. \tag{12.8}$$

By the independence of h and w and the fact that g, h and w all have the same distribution,

$$\dim_g(A) = \frac{(\mathbb{E}^h(\max_{i \in A} w_i))^2}{m^2}.$$

Therefore,

$$\left|\dim_g(A) - (1-\alpha^2)\right| \leq 2\left|(\dim_g(A))^{1/2} - \sqrt{1-\alpha^2}\right|$$

$$\leq \frac{2}{m}\left|\mathbb{E}^h\left(\max_{i \in A} w_i\right) - \sqrt{1-\alpha^2}m\right|.$$

Letting $x = 4\delta/\sqrt{1-\alpha^2}$ finishes the proof. □

3 Dimension of Near-Maximal Sets

As usual, let $g = (g_1, \ldots, g_n)$ be a Gaussian random vector with $\mathbb{E}(g_i) = 0$ and $\mathbb{E}(g_i^2) = 1$ for all i. For each i and $\delta > 0$, let $B(i, \delta) := \{j : \mathbb{E}(g_i g_j) > 1 - \delta\}$, and define

$$V(\delta) := \max_i |B(i, \delta)|. \tag{12.9}$$

Note that since $\mathbb{E}(g_i^2) = 1$, it follows that $V(\delta) \geq 1$ for any $\delta > 0$. Let

$$m := \mathbb{E}\left(\max_i g_i\right).$$

For any $\epsilon \in (0, 1)$, let

$$A(\epsilon) := \{i : g_i \geq (1-\epsilon)m\}. \tag{12.10}$$

The object of interest in this section is the size of $A(\epsilon)$ for small ϵ. It is easy to show by direct computation and the inequalities (A.2) from Appendix A that for large n,

$$\mathbb{E}|A(\epsilon)| \sim n^{1-(1-\epsilon)^2 c^2},$$

where $c = m/\sqrt{2\log n}$. By Markov's inequality, this gives an upper bound on the size of $A(\epsilon)$. However, unless $c \approx 1$, one cannot expect that this expected value reflects the true size of $A(\epsilon)$, which should be much smaller. Indeed, one can expect that in general the 'true behavior' of $|A(\epsilon)|$ should be like

$$|A(\epsilon)| \sim n^{f(\epsilon)}$$

for some function f such that $\lim_{\epsilon \to 0} f(\epsilon) = 0$. The following result can be used to prove this under pretty general conditions. Examples will follow.

Theorem 12.4 *Under the above setup, for any $\epsilon \in (0, 1)$ we have the bound*

$$\mathbb{E}\big(\log|A(\epsilon)|;\, A(\epsilon) \ne \emptyset\big) \le \inf_{\delta \in (0,1)} \left(\log V(\delta) + \frac{C \max\{\epsilon m^2, m\}}{\delta} \right),$$

where C is a universal constant.

Proof Throughout the proof, C will denote arbitrary universal constants.
For every $A \subseteq \{1, \ldots, n\}$, define

$$m(A) := \mathbb{E}\Big(\max_{i \in A} g_i\Big).$$

In particular, $m(\emptyset) = -\infty$. Clearly, if $A \subseteq B$, then $m(A) \le m(B)$. Moreover, whenever $A \ne \emptyset$, it is easy to see that $m(A) \ge 0$. From Proposition 12.2 it follows that

$$\log|A| \le \inf_{\delta \in (0,1)} \left(\log V(\delta) + C \frac{m(A)^2}{\delta} \right). \tag{12.11}$$

Fix $\epsilon \in (0, 1)$. Let g' be an independent copy of g and define

$$h := (1 - \epsilon)g + \sqrt{2\epsilon - \epsilon^2}\, g'.$$

Then h has the same law as g. Let

$$M_\epsilon^g := \max_{i \in A(\epsilon)} g_i, \qquad M_\epsilon^{g'} := \max_{i \in A(\epsilon)} g_i', \qquad M_\epsilon^h := \max_{i \in A(\epsilon)} h_i.$$

(We follow the usual convention that the maximum of an empty set is $-\infty$.) Since $g_i \ge (1 - \epsilon)m$ for all $i \in A(\epsilon)$, we have

$$M_\epsilon^h \ge (1 - \epsilon)^2 m + \sqrt{2\epsilon - \epsilon^2}\, M_\epsilon^{g'}.$$

Written differently, this gives

$$M_\epsilon^{g'} \le \sqrt{2\epsilon - \epsilon^2}\, m + \frac{M_\epsilon^h - m}{\sqrt{2\epsilon - \epsilon^2}} \le \sqrt{2\epsilon - \epsilon^2}\, m + \frac{\max_i h_i - m}{\sqrt{2\epsilon - \epsilon^2}}.$$

3 Dimension of Near-Maximal Sets

Thus, by inequality (A.7) of Appendix A, we get that for any $t \geq 0$,

$$\mathbb{P}\big(M_\epsilon^{g'} \geq \sqrt{2\epsilon - \epsilon^2}\, m + t\big)$$
$$\leq \mathbb{P}\Big(\max_i h_i - m \geq t\sqrt{2\epsilon - \epsilon^2}\Big) \leq \exp\left(-\frac{(2\epsilon - \epsilon^2)t^2}{2}\right). \tag{12.12}$$

Now let $\mathbb{P}^g$ and $\mathbb{E}^g$ denote the conditional probability and conditional expectation given g. Note that

$$\big\{\mathbb{E}^g\big(M_\epsilon^{g'}\big) \geq \sqrt{2\epsilon - \epsilon^2}\, m + 2t\big\}$$
$$\subseteq \big\{M_\epsilon^{g'} \geq \sqrt{2\epsilon - \epsilon^2}\, m + t\big\} \cup \big\{M_\epsilon^{g'} - \mathbb{E}^g\big(M_\epsilon^{g'}\big) \leq -t,\ A(\epsilon) \neq \emptyset\big\}.$$

Thus, from (12.12) and inequality (A.7) (and the independence of g and g'), we conclude that

$$\mathbb{P}\big(\mathbb{E}^g\big(M_\epsilon^{g'}\big) \geq \sqrt{2\epsilon - \epsilon^2}\, m + 2t\big)$$
$$\leq \exp\left(-\frac{(2\epsilon - \epsilon^2)t^2}{2}\right) + e^{-t^2/2} \leq 2\exp\left(-\frac{(2\epsilon - \epsilon^2)t^2}{2}\right).$$

Now note that $\mathbb{E}^g(M_\epsilon^{g'}) \geq 0$ whenever $A(\epsilon) \neq \emptyset$. Putting $\gamma := \sqrt{2\epsilon - \epsilon^2}\, m$, the above tail bound gives

$$\mathbb{E}\big(\big(\mathbb{E}^g\big(M_\epsilon^{g'}\big)\big)^2;\, A(\epsilon) \neq \emptyset\big) = \int_0^\infty 2u\, \mathbb{P}\big(\mathbb{E}^g\big(M_\epsilon^{g'}\big) \geq u\big)du$$
$$\leq \gamma^2 + \int_0^\infty 4(\gamma + 2t)\, \mathbb{P}\big(\mathbb{E}^g\big(M_\epsilon^{g'}\big) \geq \gamma + 2t\big)\, dt$$
$$\leq \gamma^2 + \int_0^\infty 8(\gamma + 2t)\, \exp\left(-\frac{\gamma^2 t^2}{2m^2}\right) dt$$
$$= \gamma^2 + 4\sqrt{2\pi}\, m + \frac{16m^2}{\gamma^2}.$$

When $\epsilon \geq 1/m$, we have

$$\gamma^2 + 4\sqrt{2\pi}\, m + \frac{16m^2}{\gamma^2} \leq 2\epsilon m^2 + 4\sqrt{2\pi}\, m + \frac{16}{\epsilon}$$
$$\leq \big(2 + 4\sqrt{2\pi} + 16\big)\epsilon m^2 \leq 30\epsilon m^2.$$

On the other hand, $\mathbb{E}^g(M_\epsilon^{g'})$ is an increasing function of ϵ, and as noted before, non-negative whenever $A(\epsilon) \neq \emptyset$. Thus, when $\epsilon < 1/m$,

$$\mathbb{E}\big(\big(\mathbb{E}^g\big(M_\epsilon^{g'}\big)\big)^2;\, A(\epsilon) \neq \emptyset\big) \leq \mathbb{E}\big(\big(\mathbb{E}^g\big(M_{1/m}^{g'}\big)\big)^2 \mathbb{I}_{A(1/m)\neq\emptyset}\big) \leq 30m.$$

Combining, we see that for every $\epsilon \in (0, 1)$,
$$\mathbb{E}\big((\mathbb{E}^g(M_\epsilon^{g'}))^2; A(\epsilon) \neq \emptyset\big) \leq 30 \max\{\epsilon m^2, m\}.$$
By the independence of g and g', we have the crucial identity
$$\mathbb{E}^g(M_\epsilon^{g'}) \equiv m(A(\epsilon)).$$
Using (12.11) and the last two equations, we get that for any $\delta \in (0, 1)$,
$$\mathbb{E}(\log|A(\epsilon)|; A(\epsilon) \neq \emptyset) \leq \mathbb{E}\left(\log V(\delta) + \frac{Cm(A(\epsilon))^2}{\delta}\mathbb{I}_{A(\epsilon) \neq \emptyset}\right)$$
$$\leq \log V(\delta) + \frac{C \max\{\epsilon m^2, m\}}{\delta}.$$
This completes the proof. □

4 Applications

We will now apply Theorem 12.4 to two familiar examples. First, consider the Sherrington-Kirkpatrick model of spin glasses, defined in Chap. 1, Sect. 1.3. Let n be the number of particles and H_n be the Hamiltonian. Define
$$A(\epsilon) := \left\{\sigma \in \{-1, 1\}^n : H_n(\sigma) \leq (1-\epsilon)\mathbb{E}\left(\min_{\sigma'} H_n(\sigma')\right)\right\}.$$
We wish to show that if $\epsilon = o(1)$ as $n \to \infty$, then
$$|A(\epsilon)| = 2^{o(n)}.$$
The following proposition is proved using Theorem 12.4.

Proposition 12.5 *In the SK model,*
$$\mathbb{E}(\log|A(\epsilon)|; A(\epsilon) \neq \emptyset) \leq \begin{cases} Cn\sqrt{\epsilon \log(1/\epsilon)} & \text{if } \epsilon \geq n^{-1/2}, \\ Cn^{3/4}\sqrt{\log n} & \text{if } \epsilon < n^{-1/2}, \end{cases}$$
where C is a universal constant.

Proof Let
$$g_\sigma := -\frac{H_n(\sigma)}{\sqrt{(n-1)/2}}$$
so that $\mathrm{Var}(g_\sigma) = 1$ for all σ and
$$A(\epsilon) = \left\{\sigma : g_\sigma \geq (1-\epsilon)\mathbb{E}\left(\max_{\sigma'} g_{\sigma'}\right)\right\}.$$

4 Applications

This brings us into the setting of Theorem 12.4. Note that

$$\mathbb{E}(g_\sigma g_\tau) = \frac{(\sigma \cdot \tau)^2}{n(n-1)} - \frac{1}{n-1}.$$

Thus, given any $\delta \in (0, 1)$ and $\sigma \in \{-1, 1\}^n$,

$$\left|\{\tau : \mathbb{E}(g_\sigma g_\tau) \geq 1 - \delta\}\right| \leq \left|\{\tau : |\sigma \cdot \tau| \geq \sqrt{1-\delta}\,(n-1)\}\right|$$

$$\leq \sum_{k=\lceil\sqrt{1-\delta}\,(n-1)\rceil}^{n} \binom{n}{\frac{n+k}{2}}.$$

Let $k = (1-a)n$, where a is some constant in $(0, 1)$. By Stirling's approximation, we have

$$\log\binom{n}{\frac{n+k}{2}} \sim n\log n - \frac{n+k}{2}\log\frac{n+k}{2} - \frac{n-k}{2}\log\frac{n-k}{2}$$

$$\sim -\frac{n}{2}\left((2-a)\log\frac{2-a}{2} + a\log\frac{a}{2}\right).$$

When a is close to zero, the dominant term inside the brackets is $a \log a$. Since $\sqrt{1-\delta} \sim 1 - \frac{\delta}{2}$, this gives us that for some universal constant C,

$$\log V(\delta) \leq -Cn\delta \log \delta,$$

where V is defined as in (12.9). Since $\mathbb{E}(\max_\sigma g_\sigma) = c\sqrt{n}$ for some constant c that stays bounded away from 0 and 1 as $n \to \infty$ (this follows from the results of Appendix A), we can now use Theorem 12.4 to conclude that

$$\mathbb{E}\big(\log|A(\epsilon)|; A(\epsilon) \neq \emptyset\big) \leq C \inf_\delta \left(-n\delta\log\delta + \frac{\max\{\epsilon n, \sqrt{n}\}}{\delta}\right).$$

The proof is now completed by taking

$$\delta = \sqrt{\frac{\epsilon}{\log(1/\epsilon)}}$$

if $\epsilon \geq n^{-1/2}$ and $\delta = n^{-1/4}(\log n)^{-1/2}$ when $\epsilon < n^{-1/2}$. □

Next consider the $(1+1)$-dimensional Gaussian random polymer model defined in Chap. 1, Sect. 1.2. Let $H_n(p)$ be the energy of a path p of length n. Let $\mathcal{P}_n$ denote the set of all paths of length n, and define

$$A(\epsilon) = \left\{p \in \mathcal{P}_n : H_n(p) \leq (1-\epsilon)\mathbb{E}\left(\min_{p' \in \mathcal{P}_n} H_n(p')\right)\right\}.$$

The following estimates are proved using Theorem 12.4.

Proposition 12.6 *In the last passage percolation model defined above,*

$$\mathbb{E}(\log|A(\epsilon)|; A(\epsilon) \neq \emptyset) \leq \begin{cases} Cn\sqrt{\epsilon \log(1/\epsilon)} & \text{if } \epsilon \geq n^{-1/2}, \\ Cn^{3/4}\sqrt{\log n} & \text{if } \epsilon < n^{-1/2}, \end{cases}$$

where C is a universal constant.

Proof Given $\delta \in (0, 1)$ and $p \in \mathcal{P}_n$, let us try to bound the number of paths having at least $(1 - \delta)(n + 1)$ vertex intersections with p. Consider such a path p'. Now p' 'goes along' with p for a while, and then 'leaves' p to take its own course, and again comes back to 'join' p, and so on. Let ℓ_i be the step at which p' leaves p for the ith time, and j_i be the step at which it comes back to join p after step ℓ_i. Then $\ell_1 < j_1 < \ell_2 < j_2 < \cdots$. Since p and p' intersect on at least $(1 - \delta)(n + 1)$ vertices, the number of ways to choose $\ell_1, j_1, \ldots$ is at most of the order of

$$\binom{n+1}{2\delta(n+1)} \sim \exp(-Cn\delta \log(1/\delta)).$$

Given the choice of $\ell_1, j_1, \ldots$, the number of ways to choose p' is bounded by the number of oriented paths of length $\delta(n + 1)$, which is $2^{\delta(n+1)}$. Combining, we see that if $V(\delta)$ is the maximum possible number of choices of p' (as defined in (12.9)), then

$$\log V(\delta) \leq -Cn\delta \log(1/\delta).$$

The rest of the proof proceeds as in the previous example. □

Appendix A
Gaussian Random Variables

This appendix contains some well-known results about Gaussian random variables that have been used several times in this monograph.

1 Tail Bounds

The following pair of inequalities is collectively known as the Mills ratio bounds. For a standard Gaussian random variable g, for any $x > 0$,

$$\frac{xe^{-x^2/2}}{\sqrt{2\pi}(1+x^2)} \leq \mathbb{P}(g > x) \leq \frac{e^{-x^2/2}}{x\sqrt{2\pi}}. \tag{A.1}$$

The proof is not difficult, and may be found in numerous standard texts on probability and statistics. The inequalities in the above form were probably first observed by Gordon (1941). Another useful inequality is

$$\mathbb{P}(g > x) \leq e^{-x^2/2}, \tag{A.2}$$

which follows simply by optimizing over θ in the inequality $\mathbb{P}(g > x) \leq e^{-\theta x}\mathbb{E}(e^{\theta g})$, using the formula $\mathbb{E}(e^{\theta g}) = e^{\theta^2/2}$.

2 Size of the Maximum

It is a well-known result that the maximum of n i.i.d. Gaussian random variables is asymptotically like $\sqrt{2\log n}$, with fluctuations of order $(\log n)^{-1/2}$. In fact, much finer details are known, such as the limiting distribution of the maximum under the correct centering and scaling (see e.g. Leadbetter et al. 1983).

However, what is not so well-known (but equally easy) is that the maximum of n standard Gaussian random variables, even if they are not independent, cannot be

S. Chatterjee, *Superconcentration and Related Topics*,
Springer Monographs in Mathematics, DOI 10.1007/978-3-319-03886-5,
© Springer International Publishing Switzerland 2014

much larger than $\sqrt{2\log n}$. In fact, irrespective of the covariance structure,

$$\mathbb{E}\left(\max_{1\leq i\leq n} g_i\right) \leq \sqrt{2\log n}. \tag{A.3}$$

This is easily proved as follows. Suppose $g_1, \ldots, g_n$ are standard Gaussian random variables, not necessarily independent. Then for any $\beta > 0$,

$$\mathbb{E}\left(\max_i g_i\right) = \frac{1}{\beta}\mathbb{E}\left(\log e^{\beta \max_i g_i}\right)$$

$$\leq \frac{1}{\beta}\mathbb{E}\left(\log \sum_{i=1}^n e^{\beta g_i}\right) \leq \frac{1}{\beta}\log \sum_{i=1}^n \mathbb{E}\left(e^{\beta g_i}\right) = \frac{\beta}{2} + \frac{\log n}{\beta}.$$

The claim is proved by taking $\beta = \sqrt{2\log n}$. Similarly,

$$\mathbb{E}\left(\max_i |g_i|\right) = \frac{1}{\beta}\mathbb{E}\left(\log e^{\beta \max_i |g_i|}\right)$$

$$\leq \frac{1}{\beta}\mathbb{E}\left(\log \sum_{i=1}^n e^{\beta |g_i|}\right) \leq \frac{1}{\beta}\log \sum_{i=1}^n \mathbb{E}\left(e^{\beta g_i} + e^{-\beta g_i}\right)$$

$$= \frac{\beta}{2} + \frac{\log(2n)}{\beta}.$$

Taking $\beta = \sqrt{2\log(2n)}$ gives

$$\mathbb{E}\left(\max_i |g_i|\right) \leq \sqrt{2\log(2n)}.$$

A stronger bound is given by the following lemma.

Lemma A.1 *Let $g_1, \ldots, g_n$ be i.i.d. standard Gaussian random variables. If $n \geq 2$ then for any $p \geq 1$,*

$$\mathbb{E}\left|\max_i g_i\right|^p \leq \mathbb{E}\max_i |g_i|^p \leq C(p)(\log n)^{p/2},$$

where $C(p)$ is a constant that depends only on p.

Proof Combine the inequality $\mathbb{E}(\max_i g_i) \leq \sqrt{2\log n}$ with the concentration of the maximum (see later in this Appendix), and observe that $\max |g_i|$ is the maximum of the concatenation of the vectors g and $-g$. □

Sometimes, information about the size of the maximum can be obtained from a famous comparison method due to Slepian (1962):

Lemma A.2 (Slepian's lemma) *Suppose g and h are centered Gaussian random n-vectors with $\mathbb{E}(g_i^2) = \mathbb{E}(h_i^2)$ for each i and $\mathbb{E}(g_i g_j) \geq \mathbb{E}(h_i h_j)$ for each i, j. Then for each $x \in \mathbb{R}$,*

$$\mathbb{P}\left(\max_i g_i > x\right) \leq \mathbb{P}\left(\max_i h_i > x\right).$$

In particular, $\mathbb{E} \max_i g_i \leq \mathbb{E} \max_i h_i$.

The following lemma due to Sudakov gives a simple way to compute lower bounds on the expected value of the maximum. For a proof of this result, see Talagrand (2005, Lemma 2.1.2).

Lemma A.3 Sudakov minoration *Let g be a centered Gaussian n-vector. Suppose a is a constant such that $\mathbb{E}(g_i - g_j)^2 \geq a$ for all $i \neq j$. Then*

$$\mathbb{E} \max_i g_i \geq Ca\sqrt{\log n},$$

where C is a positive universal constant.

3 Integration by Parts

Suppose g is a standard Gaussian random variable, and $f : \mathbb{R} \to \mathbb{R}$ is an absolutely continuous function. Under the assumption that $\mathbb{E}|f'(g)| < \infty$, a standard application of integration by parts gives the well-known identity

$$\mathbb{E} g f(g) = \mathbb{E} f'(g).$$

This identity can be easily generalized to n dimensions. Suppose that $g = (g_1, \ldots, g_n)$ is a centered Gaussian vector, possibly with correlations. If $f : \mathbb{R}^n \to \mathbb{R}$ is an absolutely continuous function such that $|\nabla f(g)|$ has finite expectation (where ∇f is the gradient of f and $|\cdot|$ is the Euclidean norm), then for any i,

$$\mathbb{E}(g_i f(g)) = \sum_{j=1}^n \mathbb{E}(g_i g_j) \mathbb{E}(\partial_i f(g)), \quad (A.4)$$

where $\partial_i f$ is the partial derivative of f in the ith coordinate. The above identity can be derived from the previous one by writing g as a linear transformation of a standard Gaussian random vector.

4 The Gaussian Concentration Inequality

The Gaussian Poincaré inequality gives a bound on the variance of a function of Gaussian random variables. In particular, it says that if the function is Lipschitz, then the variance is bounded by the square of the Lipschitz constant. For Lipschitz functions, it is possible to derive a much stronger version of the Poincaré inequality that gives a sub-Gaussian tail bound. If $f : \mathbb{R}^n \to \mathbb{R}$ is a Lipschitz function with Lipschitz constant K and g is an n-dimensional standard Gaussian random vector, then for any $x \geq 0$

$$\mathbb{P}(f(g) \geq \mathbb{E}f(g) + x) \leq e^{-x^2/2K^2} \quad \text{and}$$
$$\mathbb{P}(f(g) \leq \mathbb{E}f(g) - x) \leq e^{-x^2/2K^2}. \tag{A.5}$$

This is known as the Gaussian concentration inequality. To prove this inequality, assume without loss of generality that $\mathbb{E}f(g) = 0$. Take any $\theta > 0$ and let $v = e^{\theta f}$. Let $(\cdot, \cdot)$ denote the usual inner product on $L^2(\gamma^n)$, where γ^n is the n-dimensional standard Gaussian measure. By the covariance lemma (Lemma 2.1) and the representation (2.2),

$$(f, v) = \int_0^\infty e^{-t} \int \nabla v \cdot P_t \nabla f \, d\gamma^n \, dt$$
$$= \int_0^\infty e^{-t} \int \theta e^{\theta f} \nabla f \cdot P_t \nabla f \, d\gamma^n \, dt.$$

By the Cauchy-Schwarz inequality and the Lipschitzness of f, $|\nabla f \cdot P_t \nabla f| \leq K^2$ everywhere. Thus the above identity gives that for all $\theta > 0$

$$\frac{d}{d\theta}\mathbb{E}(e^{\theta f(g)}) = (f, e^{\theta f}) \leq K^2 \theta \, \mathbb{E}(e^{\theta f(g)}),$$

which, in turn, implies that

$$\mathbb{E}(e^{\theta f(g)}) \leq e^{K^2 \theta^2/2}. \tag{A.6}$$

Optimizing the inequality $\mathbb{P}(f(g) \geq x) \leq e^{-\theta x} \mathbb{E}(e^{\theta f(g)})$ over θ gives the bound $\mathbb{P}(f(g) \geq x) \leq e^{-x^2/2K^2}$. The bound for the lower tail can be obtained by replacing f with $-f$.

5 Concentration of the Maximum

Let $g = (g_1, \ldots, g_n)$ be a Gaussian random vector, whose coordinates are not necessarily independent. One can express g as Ah, where h is a standard Gaussian vector and A is a matrix such that AA^T is the covariance matrix of g. Let a_i denote the ith row of A and a_{ij} denote the jth element of a_i. Then $\max_i g_i = f(h)$, where $f : \mathbb{R}^n \to \mathbb{R}$ is the function

$$f(x) := \max_i a_i \cdot x.$$

5 Concentration of the Maximum

Take any x and let i_x be the i that maximizes $a_i \cdot x$. (We ignore the set of measure zero on which the maximum is attained at multiple indices.) Then for each j, $\partial_j f(x) = a_{i_x j}$. Thus,

$$|\nabla f(x)|^2 = \sum_j a_{i_x j}^2 = \mathrm{Var}(g_{i_x}).$$

By the concentration inequality for Lipschitz functions, this implies that

$$\mathbb{P}\left(\max_i g_i - \mathbb{E}\max_i g_i \geq t\right) \leq e^{-t^2/2\sigma^2},$$
$$\mathbb{P}\left(\max_i g_i - \mathbb{E}\max_i g_i \leq -t\right) \leq e^{-t^2/2\sigma^2}, \quad (A.7)$$

where $\sigma^2 := \max_i \mathrm{Var}(g_i)$ and $t \geq 0$. The above inequalities were proved by Tsirelson et al. (1976), although they follow (with slightly worse constants) from the earlier works of Borell (1975) and Sudakov and Cirelson [Tsirelson] (1974).

Appendix B
Hypercontractivity

This appendix gives a proof of the hypercontractive inequality for the Ornstein-Uhlenbeck semigroup. This inequality is one of the key tools used in this monograph, so I felt that the curious reader deserves to see a proof.

The route I will follow here is by now the standard argument via the logarithmic Sobolev inequality for the Gaussian measure. The logarithmic Sobolev inequality says that if γ^n is the standard Gaussian measure on $\mathbb{R}^n$, and $f : \mathbb{R}^n \to \mathbb{R}$ is an absolutely continuous function, then

$$\int f^2 \log \frac{f^2}{\int f^2 d\gamma^n} d\gamma^n \leq 2 \int |\nabla f|^2 d\gamma^n \tag{B.1}$$

whenever both sides are finite. This important property of the Gaussian measure was discovered by Gross (1975), who connected it to other properties like hypercontractivity. The logarithmic Sobolev inequality for the Gaussian measure is a consequence of the dynamical formula (2.4) from Chap. 2. The derivation goes as follows (taken from Ledoux 2001). Let $v := f^2$. Let P_t be the n-dimensional OU semigroup and let $v_t := P_t f$, so that $v_t \to \int v d\gamma$ pointwise. Let (u, w) denote the inner product of two functions u and w in $L^2(\gamma^n)$. Under mild regularity conditions on v, a simple computation gives

$$\int f^2 \log \frac{f^2}{\int f^2 d\gamma^n} d\gamma^n = \int v \log v \, d\gamma^n - \int v \, d\gamma^n \log \int v \, d\gamma^n$$

$$= -\int_0^\infty \frac{\partial}{\partial t}(v_t, \log v_t) \, dt$$

$$= -\int_0^\infty \left(\frac{\partial v_t}{\partial t}, 1 + \log v_t \right) dt.$$

By the heat equation, $\partial v_t/\partial t = L v_t$ where L is the OU generator. By the formula (2.2) from Chap. 2,

$$(Lv_t, 1 + \log v_t) = -\int \frac{|\nabla v_t|^2}{v_t} d\gamma^n.$$

An application of the Cauchy-Schwarz inequality and the identity $\nabla v_t = e^{-t} P_t \nabla v$ (as observed in formula (2.3) of Chap. 2) shows that

$$|\nabla v_t|^2 = e^{-2t} |P_t \nabla v|^2 \le e^{-2t} v_t P_t \left(\frac{|\nabla v|^2}{v} \right).$$

Combining the last three displays gives

$$\int f^2 \log \frac{f^2}{\int f^2 d\gamma} d\gamma \le \int_0^\infty e^{-2t} \int P_t \left(\frac{|\nabla v|^2}{v} \right) d\gamma \, dt$$

$$= \int_0^\infty e^{-2t} \int \frac{|\nabla v|^2}{v} d\gamma \, dt = \frac{1}{2} \int \frac{|\nabla v|^2}{v} d\gamma.$$

Since $\nabla v = 2 f \nabla f$, this completes the proof of (B.1).

We are now ready to prove the hypercontractive inequality for the OU semigroup. Recall that the hypercontractive inequality for the OU semigroup says that for any $p > 1$, any $t \ge 0$, and any $f \in L^p(\gamma^n)$,

$$\|P_t f\|_{L^{q(t)}(\gamma^n)} \le \|f\|_{L^p(\gamma^n)}, \tag{B.2}$$

where

$$q(t) := 1 + (p-1) e^{2t}.$$

Note that $q(t) > p$ if $t > 0$. The hypercontractive inequality for the OU semigroup was discovered by Nelson (1973) and explored in connection with the logarithmic Sobolev inequality by Gross.

We will now derive the hypercontractive inequality as a consequence of the logarithmic Sobolev inequality (following the exposition in Guionnet and Zegarlinski 2003). Take any f. Let $v := |f|$. Since $|P_t f| \le P_t v$ and $\|f\|_{L^p(\gamma^n)} = \|v\|_{L^p(\gamma^n)}$, it suffices to prove (B.2) only for non-negative f. So assume $f \ge 0$ everywhere. Let $f_t := P_t f$ and $r(t) := \int f_t^{q(t)} d\gamma^n$. Then by the heat equation and the formula (2.2) from Chap. 2,

$$r'(t) = \int f_t^{q(t)} \frac{\partial}{\partial t} (q(t) \log f_t) d\gamma^n$$

$$= q'(t) \int f_t^{q(t)} \log f_t \, d\gamma^n + q(t) \int f_t^{q(t)-1} \frac{\partial f_t}{\partial t} d\gamma^n$$

$$= q'(t) \int f_t^{q(t)} \log f_t \, d\gamma^n + q(t) \int f_t^{q(t)-1} L f_t \, d\gamma^n$$

$$= \frac{q'(t)}{q(t)} \int f_t^{q(t)} \log f_t^{q(t)} d\gamma^n - q(t)\big(q(t) - 1\big) \int f_t^{q(t)-2} |\nabla f_t|^2 d\gamma^n.$$

Further note that $q'(t) = 2(q(t) - 1)$. Therefore,

$$\frac{\partial}{\partial t} \log \|f_t\|_{L^{q(t)}(\gamma^n)}$$

$$= \frac{\partial}{\partial t} \frac{\log r(t)}{q(t)} = -\frac{q'(t) \log r(t)}{q(t)^2} + \frac{r'(t)}{q(t) r(t)}$$

$$= \frac{q'(t)}{q(t)^2 r(t)} \int f_t^{q(t)} \log \frac{f_t^{q(t)}}{r(t)} d\gamma^n - \frac{q(t) - 1}{r(t)} \int f_t^{q(t)-2} |\nabla f_t|^2 d\gamma^n$$

$$= \frac{q'(t)}{q(t)^2 r(t)} \left(\int f_t^{q(t)} \log \frac{f_t^{q(t)}}{r(t)} d\gamma^n - \frac{q(t)^2}{2} \int f_t^{q(t)-2} |\nabla f_t|^2 d\gamma^n \right).$$

The logarithmic Sobolev inequality (B.1) applied to the function $f_t^{q(t)/2}$ gives

$$\int f_t^{q(t)} \log \frac{f_t^{q(t)}}{r(t)} d\gamma^n \leq \frac{q(t)^2}{2} \int f_t^{q(t)-2} |\nabla f_t|^2 d\gamma^n.$$

Thus, $\frac{\partial}{\partial t} \log \|f_t\|_{L^{q(t)}(\gamma^n)} \leq 0$ for all t. This proves (B.2).

References

Adler, R.J., Taylor, J.E.: Random Fields and Geometry. Springer Monographs in Mathematics. Springer, New York (2007)

Aizenman, M., Lebowitz, J.L., Ruelle, D.: Some rigorous results on the Sherrington-Kirkpatrick spin glass model. Commun. Math. Phys. **112**(1), 3–20 (1987)

Aldous, D.J.: The $\zeta(2)$ limit in the random assignment problem. Random Struct. Algorithms **18**(4), 381–418 (2001)

Aldous, D., Diaconis, P.: Longest increasing subsequences: from patience sorting to the Baik-Deift-Johansson theorem. Bull. Am. Math. Soc. **36**(4), 413–432 (1999)

Aldous, D.J., Bordenave, C., Lelarge, M.: Near-minimal spanning trees: a scaling exponent in probability models. Ann. Inst. Henri Poincaré Probab. Stat. **44**(5), 962–976 (2008)

Aldous, D.J., Bordenave, C., Lelarge, M.: Dynamic programming optimization over random data: the scaling exponent for near optimal solutions. SIAM J. Comput. **38**(6), 2382–2410 (2009)

Alexander, K.S., Zygouras, N.: Subgaussian concentration and rates of convergence in directed polymers. Preprint (2012). Available at http://arxiv.org/abs/1204.1819

Aubrun, G.: A sharp small deviation inequality for the largest eigenvalue of a random matrix. In: Séminaire de Probabilités XXXVIII. Lecture Notes in Math., vol. 1857, pp. 320–337. Springer, Berlin (2005)

Baik, J., Deift, P., Johansson, K.: On the distribution of the length of the longest increasing subsequence of random permutations. J. Am. Math. Soc. **12**(4), 1119–1178 (1999)

Beckner, W.: Inequalities in Fourier analysis. Ann. Math. **102**, 159–182 (1975)

Benaïm, M., Rossignol, R.: Exponential concentration for first passage percolation through modified Poincaré inequalities. Ann. Inst. Henri Poincaré Probab. Stat. **44**(3), 544–573 (2008)

Benjamini, I., Rossignol, R.: Submean variance bound for effective resistance of random electric networks. Commun. Math. Phys. **280**(2), 445–462 (2008)

Benjamini, I., Kalai, G., Schramm, O.: Noise sensitivity of Boolean functions and applications to percolation. Publ. Math. IHÉS **1999**(90), 5–43 (2001)

Benjamini, I., Kalai, G., Schramm, O.: First passage percolation has sublinear distance variance. Ann. Probab. **31**(4), 1970–1978 (2003)

Berman, S.M.: Limit theorems for the maximum term in stationary sequences. Ann. Math. Stat. **35**(2), 502–516 (1964)

Biggins, J.D.: Chernoff's theorem in the branching random walk. J. Appl. Probab. **14**(3), 630–636 (1977)

Bolthausen, E., Deuschel, J.-D., Giacomin, G.: Entropic repulsion and the maximum of the two-dimensional harmonic crystal. Ann. Probab. **29**(4), 1670–1692 (2001)

Bonami, A.: Etude des coefficients Fourier des fonctiones de $L^p(G)$. Ann. Inst. Fourier (Grenoble) **20**, 335–402 (1970)

Borell, C.: The Brunn-Minkowski inequality in Gauss space. Invent. Math. **30**, 205–216 (1975)

Borodin, A., Corwin, I., Ferrari, P.: Free energy fluctuations for directed polymers in random media in 1 + 1 dimension. Preprint (2012). Available at http://arxiv.org/abs/1204.1024

Boucheron, S., Lugosi, G., Massart, S.: Concentration inequalities using the entropy method. Ann. Probab. **31**(3), 1583–1614 (2003)

Boucheron, S., Lugosi, G., Massart, P.: Concentration Inequalities: A Nonasymptotic Theory of Independence. Oxford University Press, London (2013)

Bramson, M., Zeitouni, O.: Tightness for a family of recursion equations. Ann. Probab. **37**(2), 615–653 (2009)

Bramson, M., Zeitouni, O.: Tightness of the recentered maximum of the two-dimensional discrete Gaussian free field. Commun. Pure Appl. Math. **65**, 1–20 (2011)

Bramson, M., Ding, J., Zeitouni, O.: Convergence in law of the maximum of the two-dimensional discrete Gaussian free field. Preprint (2013). Available at http://arxiv.org/abs/1301.6669

Bray, A.J., Moore, M.A.: Chaotic nature of the spin-glass phase. Phys. Rev. Lett. **58**(1), 57–60 (1987)

Chatterjee, S.: A new method of normal approximation. Ann. Probab. **36**(4), 1584–1610 (2008a)

Chatterjee, S.: Chaos, concentration, and multiple valleys. Preprint (2008b). Available at http://arxiv.org/abs/0810.4221

Chatterjee, S.: Disorder chaos and multiple valleys in spin glasses. Preprint (2009). Available at http://arxiv.org/abs/0907.3381

Chen, L.H.Y.: An inequality for the multivariate normal distribution. J. Multivar. Anal. **12**, 306–315 (1982)

Chen, W.-K.: Disorder chaos in the Sherrington-Kirkpatrick model with external field. Preprint (2011). Available at http://arxiv.org/abs/1109.3249

Chen, W.-K., Panchenko, D.: An approach to chaos in some mixed p-spin models. Preprint (2012). Available at http://arxiv.org/abs/1201.2198

Chernoff, H.: A note on an inequality involving the normal distribution. Ann. Probab. **9**, 533–535 (1981)

da Silveira, R.A., Bouchaud, J.-P.: Temperature and disorder chaos in low dimensional directed paths. Phys. Rev. Lett. **93**(1), 015901 (2004)

Daviaud, O.: Extremes of the discrete two-dimensional Gaussian free field. Ann. Probab. **34**(3), 962–986 (2006)

Dekking, F.M., Host, B.: Limit distributions for minimal displacement of branching random walks. Probab. Theory Relat. Fields **90**(3), 403–426 (1991)

Derrida, B.: Random energy model: limit of a family of disordered models. Phys. Rev. Lett. **45**, 79–82 (1980)

Derrida, B.: Random energy model: an exactly solvable model of disordered systems. Phys. Rev. B **24**, 2613–2626 (1981)

Deuschel, J.-D., Giacomin, G., Ioffe, D.: Large deviations and concentration properties for $\nabla\varphi$ interface models. Probab. Theory Relat. Fields **117**(1), 49–111 (2000)

Ding, J.: Exponential and double exponential tails for maximum of two-dimensional discrete Gaussian free field. Preprint (2011). Available at http://arxiv.org/abs/1105.5833

Ding, J., Zeitouni, O.: Extreme values for two-dimensional discrete Gaussian free field. Preprint (2012). Available at http://arxiv.org/abs/1206.0346

Durrett, R., Limic, V.: Rigorous results for the *NK* model. Ann. Probab. **31**(4), 1713–1753 (2003)

Edwards, S.F., Anderson, P.W.: Theory of spin glasses. J. Phys. F **5**, 965–974 (1975)

Efron, B., Stein, C.: The jackknife estimate of variance. Ann. Stat. **9**(3), 586–596 (1981)

Evans, S.N., Steinsaltz, D.: Estimating some features of *NK* fitness landscapes. Ann. Appl. Probab. **12**(4), 1299–1321 (2002)

Feller, W.: An Introduction to Probability Theory and Its Applications, vol. II, 2nd edn. Wiley, New York (1971)

Fisher, D.S., Huse, D.A.: Ordered phase of short-range Ising spin glasses. Phys. Rev. Lett. **56**(15), 1601–1604 (1986)

Friedgut, E.: Boolean functions with low average sensitivity depend on few coordinates. Combinatorica **18**(1), 27–35 (1998)

Friedgut, E.: Sharp thresholds of graph properties, and the k-sat problem. J. Am. Math. Soc. **12**(4), 1017–1054 (1999). With an appendix by Jean Bourgain

Friedgut, E., Kalai, G.: Every monotone graph property has a sharp threshold. Proc. Am. Math. Soc. **124**(10), 2993–3002 (1996)

Garban, C., Pete, G., Schramm, O.: The Fourier spectrum of critical percolation. Acta Math. **205**(1), 19–104 (2010)

Gardner, E.: Spin glasses with p-spin interactions. Nucl. Phys. B **257**(6), 747–765 (1985)

Giacomin, G.: Anharmonic lattices, random walks and random interfaces. Recent. Res. Dev. Stat. Phys. Transw. Res. **I**, 97–118 (2000)

Gordon, R.D.: Values of Mills' ratio of area to bounding ordinate and of the normal probability integral for large values of the argument. Ann. Math. Stat. **12**, 364–366 (1941)

Graham, B.T.: Sublinear variance for directed last-passage percolation. Preprint (2010). Available at http://arxiv.org/abs/0909.1352

Grimmett, G.R., Kesten, H.: Percolation since saint-flour. Preprint (2012). Available at http://arxiv.org/abs/1207.0373

Gross, L.: Logarithmic Sobolev inequalities. Am. J. Math. **97**(4), 1061–1083 (1975)

Gross, D., Mézard, M.: The simplest spin glass. Nucl. Phys. B **240**, 431–452 (1984)

Guerra, F.: Broken replica symmetry bounds in the mean field spin glass model. Commun. Math. Phys. **233**, 1–12 (2003)

Guerra, F., Toninelli, F.L.: The thermodynamic limit in mean field spin glass models. Commun. Math. Phys. **230**, 71–79 (2002)

Guerra, F.: Fluctuations and thermodynamic variables in mean field spin glass models. In: Albeverio, S., et al. (eds.) Stochastic Processes, Physics and Geometry. World Scientific, Singapore (1995)

Guionnet, A., Zegarlinski, B.: Lectures on logarithmic Sobolev inequalities. In: Séminaire de Probabilités, XXXVI. Lecture Notes in Math., vol. 1801, pp. 1–134. Springer, Berlin (2003)

Hammersley, J.M., Welsh, D.J.A.: First-passage percolation, subadditive processes, stochastic networks, and generalized renewal theory. In: Proc. Internat. Res. Semin., Statist. Lab., Univ. California, Berkeley, Calif. Springer, New York (1965)

Hoeffding, W.: Probability inequalities for sums of bounded random variables. J. Am. Stat. Assoc. **58**, 13–30 (1963)

Houdré, C.: Some applications of covariance identities and inequalities to functions of multivariate normal variables. J. Am. Stat. Assoc. **90**(431), 965–968 (1995)

Houdré, C., Kagan, A.: Variance inequalities for functions of Gaussian variables. J. Theor. Probab. **8**, 23–30 (1995)

Hu, X., Miller, J., Peres, Y.: Thick points of the Gaussian free field. Ann. Probab. **38**(2), 896–926 (2010)

Huse, D.A., Henley, C.L., Fisher, D.S.: Huse, Henley and Fisher respond. Phys. Rev. Lett. **55**(26), 2924 (1985)

Imbrie, J.Z., Spencer, T.: Diffusion of directed polymers in a random environment. J. Stat. Phys. **52**(3–4), 609–626 (1988)

Johansson, K.: Shape fluctuations and random matrices. Commun. Math. Phys. **209**(2), 437–476 (2000)

Kahn, J., Kalai, G., Linial, N.: The influence of variables on Boolean functions. In: FOCS 1988, pp. 68–80 (1988)

Kauffman, S.A., Levin, S.A.: Towards a general theory of adaptive walks on rugged landscapes. J. Theor. Biol. **128**, 11–45 (1987)

Kesten, H.: Aspects of first passage percolation. In: Lecture Notes in Mathematics, vol. 1180, pp. 125–264. Springer, Berlin (1986)

Kesten, H.: On the speed of convergence in first-passage percolation. Ann. Appl. Probab. **3**(2), 296–338 (1993)

Kim, J.H.: On increasing subsequences of random permutations. J. Comb. Theory, Ser. A **76**(1), 148–155 (1996)

Krzakała, F., Bouchaud, J.-P.: Disorder chaos in spin glasses. Europhys. Lett. **72**(3), 472–478 (2005)
Leadbetter, M.R., Lindgren, G., Rootzén, H.: Extremes and Related Properties of Random Sequences and Processes. Springer, New York (1983)
Ledoux, M.: The Concentration of Measure Phenomenon. Am. Math. Soc., Providence (2001)
Ledoux, M.: A remark on hypercontractivity and tail inequalities for the largest eigenvalues of random matrices. In: Séminaire de Probabilités XXXVII. Lecture Notes in Math., vol. 1832, pp. 360–369. Springer, Berlin (2003)
Ledoux, M.: Deviation inequalities on largest eigenvalues. In: Geometric Aspects of Functional Analysis, pp. 167–219. Springer, Berlin (2007)
Limic, V., Pemantle, R.: More rigorous results on the Kauffman-Levin model of evolution. Ann. Probab. **32**(3A), 2149–2178 (2004)
Matic, I., Nolen, J.: A sublinear variance bound for solutions of a random Hamilton Jacobi equation. Preprint (2012). Available at http://arxiv.org/abs/1206.2937
Mehta, M.L.: Random Matrices, 2nd edn. Academic Press, Boston (1991)
Mézard, M.: On the glassy nature of random directed polymers in two dimensions. J. Phys. Fr. **51**, 1831–1846 (1990)
Mittal, Y., Ylvisaker, D.: Limit distributions for the maxima of stationary Gaussian processes. Stoch. Process. Appl. **3**, 1–18 (1975)
Mossel, E., O'Donnell, R., Servedio, R.A.: Learning juntas. In: STOC 2003, pp. 206–212 (2003)
Mossel, E., O'Donnell, R., Oleszkiewicz, K.: Noise stability of functions with low influences: invariance and optimality. Ann. Math. (2) **171**(1), 295–341 (2010)
Nash, J.: Continuity of solutions of parabolic and elliptic equations. Am. J. Math. **80**(4), 931–954 (1958)
Nelson, E.: The free Markoff field. J. Funct. Anal. **12**, 211–227 (1973)
Newman, C.M., Piza, M.S.T.: Divergence of shape fluctuations in two dimensions. Ann. Probab. **23**(3), 977–1005 (1995)
Newman, C.M., Stein, D.L.: Short-range spin glasses: results and speculations. In: Spin Glasses. Lecture Notes in Math., vol. 1900, pp. 159–175. Springer, Berlin (2007)
Nourdin, I., Peccati, G.: Stein's method on Wiener chaos. Probab. Theory Relat. Fields **145**(1–2), 75–118 (2009)
Nourdin, I., Peccati, G.: Normal Approximations with Malliavin Calculus: From Stein's Method to Universality. Cambridge University Press, Cambridge (2012)
Nourdin, I., Viens, F.G.: Density formula and concentration inequalities with Malliavin calculus. Electron. J. Probab. **14**(78), 2287–2309 (2009)
Nualart, D.: Malliavin Calculus and Its Applications. American Mathematical Society, Providence (2009)
Panchenko, D.: On differentiability of the Parisi formula. Electron. Commun. Probab. **13**, 241–247 (2008)
Panchenko, D.: The Parisi ultrametricity conjecture. Ann. Math. (2) **177**(1), 383–393 (2013a)
Panchenko, D.: The Sherrington-Kirkpatrick Model. Springer, Berlin (2013b)
Panchenko, D., Talagrand, M.: On the overlap in the multiple spherical SK models. Ann. Probab. **35**(6), 2321–2355 (2007)
Parisi, G., Rizzo, T.: Phase diagram and large deviations in the free-energy of mean-field spin-glasses. Phys. Rev. B **79**, 134205 (2009)
Pemantle, R., Peres, Y.: Planar first-passage percolation times are not tight. In: Probability and Phase Transition. NATO Adv. Sci. Inst. Ser. C Math. Phys. Sci., vol. 420, pp. 261–264. Kluwer Academic, Dordrecht (1994)
Pickands, J., III: Maxima of stationary Gaussian processes. Z. Wahrscheinlichkeitstheor. Verw. Geb. **7**, 190–223 (1967)
Pickands, J., III: Upcrossing probabilities for stationary Gaussian processes. Trans. Am. Math. Soc. **145**, 51–73 (1969a)
Pickands, J., III: Asymptotic properties of the maximum in a stationary Gaussian process. Trans. Am. Math. Soc. **145**, 75–86 (1969b)

References

Rizzo, T.: Chaos in mean-field spin-glass models. In: de Monvel, A.B., Bovier, A. (eds.) Spin Glasses: Statics and Dynamics. Progress in Probability, vol. 62, pp. 143–157 (2009)

Russo, L.: On the critical percolation probabilities. Z. Wahrscheinlichkeitstheor. Verw. Geb. **56**(2), 229–237 (1981)

Sheffield, S.: Gaussian free fields for mathematicians. Probab. Theory Relat. Fields **139**(3–4), 521–541 (2007)

Sherrington, D., Kirkpatrick, S.: Solvable model of a spin glass. Phys. Rev. Lett. **35**, 1792–1796 (1975)

Slepian, D.: The one-sided barrier problem for Gaussian noise. Bell Syst. Tech. J. **41**, 463–501 (1962)

Steele, J.M.: An Efron-Stein inequality for nonsymmetric statistics. Ann. Stat. **14**, 753–758 (1986)

Sudakov, V.N., Cirelson [Tsirelson], B.S.: Extremal properties of half-spaces for spherically invariant measures. (Russian) Problems in the theory of probability distributions, II. Zap. Nauč. Semin. Leningrad. Otdel. Mat. Inst. Steklov. (LOMI) **41**, 14–24, 165 (1974)

Talagrand, M.: On Russo's approximate zero-one law. Ann. Probab. **22**, 1576–1587 (1994)

Talagrand, M.: Concentration of measure and isoperimetric inequalities in product spaces. Publ. Math. IHÉS **81**, 73–205 (1995)

Talagrand, M.: On boundaries and influences. Combinatorica **17**(2), 275–285 (1997)

Talagrand, M.: Spin Glasses: A Challenge for Mathematicians. Cavity and Mean Field Models. Springer, Berlin (2003)

Talagrand, M.: The Generic Chaining. Upper and Lower Bounds of Stochastic Processes. Springer, Berlin (2005)

Talagrand, M.: The Parisi formula. Ann. Math. (2) **163**(1), 221–263 (2006)

Talagrand, M.: Mean Field Models for Spin Glasses, vol. I. Springer, Berlin (2011)

Talagrand, M.: Mean Field Models for Spin Glasses, vol. II. Springer, Berlin (2012)

Tracy, C.A., Widom, H.: Level-spacing distributions and the Airy kernel. Commun. Math. Phys. **159**(1), 151–174 (1994a)

Tracy, C.A., Widom, H.: Fredholm determinants, differential equations and matrix models. Commun. Math. Phys. **163**(1), 33–72 (1994b)

Tracy, C.A., Widom, H.: On orthogonal and symplectic matrix ensembles. Commun. Math. Phys. **177**(3), 727–754 (1996)

Tsirelson, B.S., Ibragimov, I.A., Sudakov, V.N.: Norms of Gaussian sample functions. In: Proceedings of the Third Japan-USSR Symposium on Probability Theory, Tashkent, 1975, vol. 550, pp. 20–41 (1976)

van den Berg, J., Kiss, D.: Sublinearity of the travel-time variance for dependent first-passage percolation. Ann. Probab. **40**(2), 743–764 (2012)

Wehr, J., Aizenman, M.: Fluctuations of extensive functions of quenched random couplings. J. Stat. Phys. **60**(3–4), 287–306 (1990)

Zhang, Y.-C.: Ground state instability of a random system. Phys. Rev. Lett. **59**(19), 2125–2128 (1987)

Author Index

A
Adler, R.J., 8, 100
Aizenman, M., 5, 9, 118
Aldous, D.J., 11, 12, 46
Alexander, K.S., 46
Anderson, P.W., 118
Aubrun, G., 29, 30, 46

B
Baik, J., 46
Beckner, W., 65
Benaïm, M., 2, 46, 51
Benjamini, I., 2–4, 6, 25, 46, 51, 54
Berman, S.M., 100, 101
Biggins, J.D., 48
Bolthausen, E., 84
Bonami, A., 65
Bordenave, C., 12
Borell, C., 141
Borodin, A., 3
Bouchaud, J.-P., 8, 9
Boucheron, S., 1, 65
Bramson, M., 48, 74, 85
Bray, A.J., 9, 120

C
Chatterjee, S., 1, 3–5, 7–14, 24, 27, 33, 46, 54, 66, 67, 73, 74, 84, 85, 92, 95, 101, 105, 118, 120, 121
Chen, L.H.Y., 6
Chen, W.-K., 9
Chernoff, H., 6
Corwin, I., 3

D
Da Silveira, R.A., 8
Daviaud, O., 126

D (cont.)
Deift, P., 46
Dekking, F.M., 46
Derrida, B., 112
Deuschel, J.-D., 8, 84
Diaconis, P., 46
Ding, J., 74, 85
Durrett, R., 41

E
Edwards, S.F., 118
Efron, B., 64
Evans, S.N., 41

F
Feller, W., 107
Ferrari, P., 3
Fisher, D.S., 7, 9, 120
Friedgut, E., 25

G
Garban, C., 25
Gardner, E., 80
Giacomin, G., 8, 83, 84
Gordon, R.D., 137
Graham, B.T., 4, 8, 14, 46, 54
Grimmett, G.R., 2
Gross, D., 80
Gross, L., 45, 143
Guerra, F., 5, 9
Guionnet, A., 15, 144

H
Hammersley, J.M., 2
Henley, C.L., 7
Hoeffding, W., 62
Host, B., 46
Houdré, C., 6

Hu, X., 126
Huse, D.A., 7, 9, 120

I
Ibragimov, I.A., 141
Imbrie, J.Z., 4
Ioffe, D., 8

J
Johansson, K., 3, 46

K
Kagan, A., 6
Kahn, J., 25, 46
Kalai, G., 2–4, 6, 25, 46, 51, 54
Kauffman, S.A., 40
Kesten, H., 2, 51
Kim, J. H., 46
Kirkpatrick, S., 5
Kiss, D., 46
Krząkała, F., 9

L
Leadbetter, M.R., 100, 137
Lebowitz, J.L., 5, 9
Ledoux, M., 1, 15, 29, 30, 46, 87, 89, 90, 143
Lelarge, M., 12
Levin, S.A., 40
Limic, V., 41
Lindgren, G., 100, 137
Linial, N., 25, 46
Lugosi, G., 1, 65

M
Massart, P., 1, 65
Matic, I., 46
Mehta, M.L., 29, 30, 89
Mezard, M., 7, 80
Miller, J., 126
Mittal, Y., 100, 101
Moore, M.A., 9, 120
Mossel, E., 25

N
Nash, J., 6
Nelson, E., 45, 144
Newman, C.M., 3, 118
Nolen, J., 46
Nourdin, I., 10, 18
Nualart, D., 18

O
O'Donnell, R., 25
Oleszkiewicz, K., 25

P
Panchenko, D., 5, 9, 29
Parisi, G., 6
Peccati, G., 18
Pemantle, R., 3, 41
Peres, Y., 3, 126
Pete, G., 25
Pickands, J., III, 100, 101
Piza, M.S.T., 3

R
Rizzo, T., 6, 9
Rootzén, H., 100, 137
Rossignol, R., 2, 46, 51
Ruelle, D., 5, 9
Russo, L., 46

S
Schramm, O., 2–4, 6, 25, 46, 51, 54
Servedio, R.A., 25
Sheffield, S., 83
Sherrington, D., 5
Slepian, D., 139
Spencer, T., 4
Steele, J.M., 64
Stein, C., 64
Stein, D.L., 118
Steinsaltz, D., 41
Sudakov, V.N., 141

T
Talagrand, M., 2, 5, 25, 29, 42, 45–47, 49, 65, 80, 90, 100, 108, 111, 113, 139
Taylor, J.E., 8, 100
Toninelli, F.L., 5
Tracy, C.A., 29, 46
Tsirelson, B.S., 141

V
Van den Berg, J., 46
Viens, F.G., 10, 18

W
Wehr, J., 118
Welsh, D.J.A., 2
Widom, H., 29, 46

Y
Ylvisaker, D., 100, 101

Z
Zegarlinski, B., 15, 144
Zeitouni, O., 48, 74, 85
Zhang, Y.-C., 7
Zygouras, N., 46

Subject Index

A
Assignment problem, 12
Asymptotic Essential Uniqueness (AEU), 11

B
Benjamini-Kalai-Schramm (BKS) theorem, 2, 4, 51, 54
Binary tree, 48
Bonami-Beckner inequality, 66
Branching random walk, 48

C
Chaos, 7, 9, 25, 33, 66, 119
    definition, 25
    equivalent to superconcentration, 9
    implies multiple valleys, 33
    in Gaussian fields, 10
    in polymers, 7, 26, 28
    in the Edwards-Anderson model, 119
    in the first eigenvector, 29
    in the generalized SK model, 112
    in the SK model, 8, 26, 28, 66
Completely monotone function, 107
Covariance lemma, 16, 18, 19, 27, 113, 120, 139

D
Definition of chaos, 25
Definition of multiple valleys, 13
Definition of superconcentration, 23
Derrida's p-spin models, 80, 111
Dimensions, 125, 129, 131
    induced, 128, 129
    of level sets, 125, 129, 131
    of near-maximal sets, 131
Dirichlet form, 16–21, 23, 26, 27, 45, 60, 63

Discrete Gaussian free field (DGFF), 73, 82–85, 94, 95, 99, 125
    on a torus, 94, 95, 99
    zero boundary condition, 82–85, 125

E
Edwards-Anderson model, 118–120
Efron-Stein inequality, 64
Equivalence of superconcentration and chaos, 9, 27
Extremal field, 73, 78, 81, 82, 84, 125, 126, 128
    levels sets, 125
    sufficient condition, 79

F
First eigenvector, 29, 30
First-passage percolation, 1–4, 45, 47, 51, 52, 54, 66
Fourier expansion, 20, 24
Free energy, 20, 26, 29, 35, 57, 58, 60, 70, 118, 120

G
Gaussian concentration inequality, 55, 90, 139
Gaussian field, 6, 10, 14, 28, 38, 73, 92, 105
Gaussian integration by parts, 17, 50, 59, 107, 115, 117, 118, 139
Gaussian Unitary Ensemble (GUE), 29, 87, 89, 90
Generalized SK model, 80, 81, 111
Generator, 15, 17, 20, 23, 48, 58, 60, 63, 144

H
Heat equation, 15, 16, 144
Hermite polynomials, 58
Hypercontractivity, 45, 65, 90, 99, 143

I

Improved Poincaré inequality, 60
Independent flips, 63, 65, 66
  chaos, 66
  hypercontractivity, 65
  semigroup, 63
Induced dimension, 128, 129
Interpolation method, 105, 108

K

Kauffman-Levin NK model, 40–43
KKL argument, 45

L

Largest eigenvalue, 29, 30, 87, 89, 90
Last-passage percolation, 3, 4, 51, 54
Level sets of Gaussian fields, 125, 129
Logarithmic Sobolev inequality, 45, 143, 144
Low correlation field, 92

M

Malliavin calculus, 18
Markov process, 15
Maxima of Gaussian fields, 6, 20, 28, 137, 140
Mills ratio, 79, 137
Monotone functions, 49, 51, 58
Multiple valleys and peaks, 11, 13, 14, 33–35, 38, 40
  definition, 13
  in Gaussian fields, 14, 38
  in polymers, 13, 34
  in the NK model, 40
  in the SK model, 13, 35

N

Near-maximal sets, 134, 135
  in polymers, 135
  in the SK model, 134
Noise-sensitivity, 24

O

Ornstein-Uhlenbeck (OU) semigroup, 17, 18, 26, 46, 58, 91, 143

P

Parisi formula, 5
Plancherel formula, 21, 58, 115
Poincaré inequality, 15, 18–23, 27, 46–48, 60, 64, 80, 139
Polymers, 3, 7, 13, 19, 26, 28, 34, 51, 54, 135

R

Random energy model (REM), 112
Random matrix, 29, 30, 87, 89, 90

S

Semigroup, 15, 18, 45, 63
Sherrington-Kirkpatrick (SK) model, 4, 8, 13, 26, 28, 35, 57, 60, 66, 80, 111, 134
Slepian's lemma, 42, 129, 138, 139
Spectral method, 57
Spherical SK model, 29
Spin glass, 4–7, 29, 74, 79, 108, 118, 134
Sudakov minoration, 40, 93, 102, 139
Sufficient condition for extremality, 79
Superconcentration, 1, 2, 4, 6, 9, 23, 28, 49, 54, 60, 73, 81, 85, 87, 89, 90, 92, 93, 95, 118
  definition, 23
  equivalent to chaos, 9
  in extremal fields, 73
  in first-passage percolation, 2, 51
  in Gaussian fields, 28, 92
  in generalized SK models, 81
  in low correlation fields, 92
  in polymers, 4, 28, 51, 54
  in subfields, 93
  in the DGFF, 85, 95
  in the Edwards-Anderson model, 118
  in the SK model, 6, 28, 60
  of largest eigenvalue, 87, 89, 90
  of monotone functions, 49

T

Talagrand's L^1–L^2 method, 43, 45, 47–49, 51, 52, 54, 57, 58, 66
Torus, 94
Tracy-Widom limit, 29

V

Variance lower bounds, 115

GPSR Compliance

The European Union's (EU) General Product Safety Regulation (GPSR) is a set of rules that requires consumer products to be safe and our obligations to ensure this.

If you have any concerns about our products, you can contact us on

ProductSafety@springernature.com

In case Publisher is established outside the EU, the EU authorized representative is:

Springer Nature Customer Service Center GmbH
Europaplatz 3
69115 Heidelberg, Germany

www.ingramcontent.com/pod-product-compliance
Ingram Content Group UK Ltd.
Pitfield, Milton Keynes, MK11 3LW, UK
UKHW021250180426
11946UKWH00003B/68